BOOK

4

NELSON MATHS

WORKBOOK

ADVANCED

YEARS 7–10

Megan Boltze
Robert Yen
Ilhea Yen

WORKSHEETS

PUZZLE SHEETS

HOMEWORK ASSIGNMENTS

Nelson Maths Workbook Book 4 Advanced
1st Edition
Megan Boltze
Robert Yen
Ilhea Yen
ISBN 9780170454568

Publisher: Robert Yen
Project editor: Alan Stewart
Editor: Anna Pang
Cover design: James Steer
Original text design by Alba Design, Adapted by: James Steer
Project designer: James Steer
Permissions researcher: Corrina Gilbert
Production controller: Karen Young
Text illustrations: Cat MacInnes
Typeset by: MPS Limited

Any URLs contained in this publication were checked for currency during the production process. Note, however, that the publisher cannot vouch for the ongoing currency of URLs.

For product information and technology assistance,
in Australia call **1300 790 853**;
in New Zealand call **0800 449 725**

For permission to use material from this text or product, please email **aust.permissions@cengage.com**

ISBN 978 0 17 045456 8

Cengage Learning Australia
Level 7, 80 Dorcas Street
South Melbourne, Victoria Australia 3205

Cengage Learning New Zealand
Unit 4B Rosedale Office Park
331 Rosedale Road, Albany, North Shore 0632, NZ

For learning solutions, visit **cengage.com.au**

Printed in China by 1010 Printing International Limited.
1 2 3 4 5 6 7 25 24 23 22 21

This 200-page workbook contains worksheets, puzzles, StartUp assignments and homework assignments written for the Australian Curriculum in Mathematics. It can be used as a valuable resource for teaching Year 10 mathematics at an advanced level, regardless of the textbook used in the classroom, and takes a holistic approach to the curriculum, including some Year 9 revision work as well. It also contains 9 10A chapters covering more complex topics, especially for students preparing for Year 11 advanced mathematics (Methods/Specialist). This workbook is designed to be handy for homework, assessment, practice, revision, relief classes or 'catch-up' lessons.

Inside:

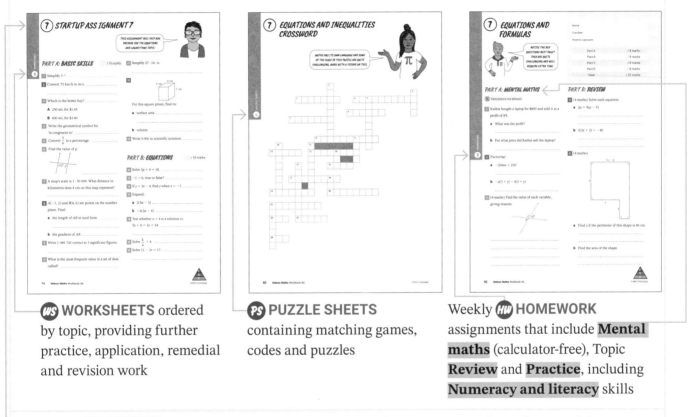

WS WORKSHEETS ordered by topic, providing further practice, application, remedial and revision work

PS PUZZLE SHEETS containing matching games, codes and puzzles

Weekly **HW HOMEWORK** assignments that include **Mental maths** (calculator-free), Topic **Review** and **Practice**, including **Numeracy and literacy** skills

StartUp assignments beginning each topic, revising skills from previous topics and prerequisite knowledge for the topic, including basic skills, review of a specific topic and a challenge problem

Word puzzles, such as a crossword or find-a-word, that reinforce the language of mathematics learned in the topic

The ideas and activities presented in this book were written by practising teachers and used successfully in the classroom.

Colour-coding of selected questions

Questions on most worksheets are graded by level of difficulty:

Complex

Standard

Foundation

CONTENTS

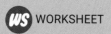 WORKSHEET   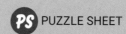 PUZZLE SHEET   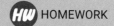 HOMEWORK

CONTENTS

MEET YOUR MATHS GUIDES ...

INTRODUCING MS LEE.

THIS WORKBOOK CONTAINS WORKSHEETS, PUZZLE SHEETS AND HOMEWORK ASSIGNMENTS

HI, I'VE BEEN TEACHING MATHS FOR OVER 20 YEARS

I BECAME GOOD AT MATHS THROUGH PRACTICE AND EFFORT

I WILL GUIDE YOU THROUGH THE WORKSHEETS

MATHS IS ABOUT MASTERING A COLLECTION OF SKILLS, AND I CAN HELP YOU DO THIS

THIS IS ZINA, A MATHS TUTOR AND MS LEE'S YEAR 12 STUDENT

HEY, I LOVE CREATING AND SOLVING PUZZLES

NOT JUST MATHS PUZZLES BUT WORD PUZZLES TOO!

PUZZLES HELP YOU THINK IN NEW AND DIFFERENT WAYS

LET ME SHOW YOU HOW, AND YOU'LL GET SMARTER ALONG THE WAY

SO LET'S GET GOING.

CURRICULUM GRID

Chapter and content	Australian curriculum strand and substrand
1 SURDS (10A)	**NUMBER AND ALGEBRA**
Surds and irrational numbers, Simplifying surds, Adding and subtracting surds, Multiplying and dividing surds, Binomial products involving surds, Rationalising the denominator	Real numbers
2 INTEREST AND DEPRECIATION	**NUMBER AND ALGEBRA**
Simple interest, Compound interest, Compound interest formula, Depreciation	Money and financial mathematics
3 GRAPHING LINES	**NUMBER AND ALGEBRA**
Length, midpoint and gradient $y = mx + c$, $ax + by + c = 0$, $y - y_1 = m(x - x_1)$, Equations of parallel and perpendicular lines	Linear and non-linear relationships
4 SURFACE AREA AND VOLUME (10A)	**MEASUREMENT AND GEOMETRY**
Surface area of prisms, cylinders, pyramids, cones, spheres, composite solids, Volumes of prisms, cylinders, pyramids, cones, spheres, composite solids, Areas of similar figures	Using units of measurement
5 PRODUCTS AND FACTORS (10A)	**NUMBER AND ALGEBRA**
The index laws, Fractional indices, Operations with algebraic fractions, Expanding binomial products, Factorising special binomial products, Factorising quadratic expressions $ax^2 + bx + c$, Factorising algebraic fractions	Real numbers Patterns and algebra Linear and non-linear relationships
6 COMPARING DATA	**STATISTICS AND PROBABILITY**
The shape of a frequency distribution, Quartiles and interquartile range, Standard deviation, Comparing means and standard deviations, Box plots, Parallel box plots, Comparing data sets, Scatterplots, Lines of best fit, Bivariate data involving time, Statistics in the media	Data representation and interpretation
7 EQUATIONS AND LOGARITHMS (10A)	**NUMBER AND ALGEBRA**
Equations with algebraic fractions, Quadratic equations $x^2 + bx + c = 0$, Equation problems, Equations and formulas, Changing the subject of a formula, Graphing inequalities on a number line, Solving inequalities, Logarithms, Logarithm laws, Exponential and logarithmic functions	Patterns and algebra Linear and non-linear relationships Real numbers
8 GRAPHING CURVES (10A)	**NUMBER AND ALGEBRA**
Direct proportion, Inverse proportion, Conversion graphs, The parabola $y = a(x - b)^2 + c$, The hyperbola $y = \dfrac{k}{x}$, The exponential curve $y = a^x$, The circle $(x - a)^2 + (y - b)^2 = r^2$	Linear and non-linear relationships Real numbers
9 FURTHER TRIGONOMETRY (10A)	**MEASUREMENT AND GEOMETRY**
Right-angled trigonometry, Bearings, Pythagoras' theorem and trigonometry in 3D, The trigonometric functions, Trigonometric equations, The sine rule, The cosine rule, The area of a triangle $A = \dfrac{1}{2} ab \sin C$	Pythagoras and trigonometry
10 SIMULTANEOUS EQUATIONS	**NUMBER AND ALGEBRA**
Solving simultaneous equations graphically, The elimination method, the substitution method, Problems involving simultaneous equations	Linear and non-linear relationships
11 QUADRATIC EQUATIONS AND THE PARABOLA (10A)	**NUMBER AND ALGEBRA**
Quadratic equations, Completing the square, the quadratic formula, Quadratic equation problems, the parabola $y = ax^2 + bx + c$, The axis of symmetry and vertex of a parabola	Linear and non-linear relationships
12 PROBABILITY	**STATISTICS AND PROBABILITY**
Relative frequency, Venn diagrams, Two-way tables, Tree diagrams, selecting with and without replacement, Dependent and independent events, Conditional probability	Chance
13 GEOMETRY	**MEASUREMENT AND GEOMETRY**
Congruent triangle proofs, Tests for quadrilaterals, Proving properties of triangles and quadrilaterals, Formal geometrical proofs, Similar figures, Finding unknown sides in similar figures, tests for similar figures, similar triangle proofs	Geometric reasoning
14 FUNCTIONS AND POLYNOMIALS (10A)	**NUMBER AND ALGEBRA**
Functions, function notation, Polynomials, Adding, subtracting and multiplying polynomials, Dividing polynomials, The remainder theorem, The factor theorem, Graphing polynomials	Patterns and algebra Linear and non-linear relationships
15 CIRCLE GEOMETRY (10A)	**MEASUREMENT AND GEOMETRY**
Parts of a circle, chord properties of circles, Angle properties of circles	Geometric reasoning

9780170454568

THIS CHAPTER IS ALL ABOUT SURDS, WHICH ARE SQUARE ROOTS THAT DON'T HAVE AN EXACT VALUE.

PART A: BASIC SKILLS / 15 marks

1 Find the median of:

3, 8, 5, 12, 8, 2 _____

2 Expand and simplify:

$2(a + 5) - (3 - a)$

_____

_____

3

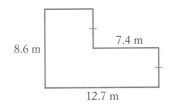

8.6 m   7.4 m   12.7 m

For this figure, calculate:

a the perimeter _____

b the area. _____

_____

4 6.2 m^2 = _____ cm^2

5 85% of what amount is $531.25? _____

6 Evaluate $\dfrac{2.5 \times 10^5}{\sqrt{15\ 625}}$. _____

7 Write an expression for the cost (in dollars) of n tickets if one ticket costs $P.

_____

8 Find the value of b in the diagram below.

150°

$b°$

_____

9 What is the size of each angle in a regular pentagon? _____

10 Solve $5x - 15 = 3x + 7$. _____

_____

_____

11 What is the probability that Deni rolls two 6s on a pair of dice? _____

12 Find the midpoint of the interval joining $(1, -4)$ and $(5, 0)$. _____

13 Is $y = -3$ parallel to the x-axis or the y-axis?

_____

14 Describe the graph of $x^2 + y^2 = 9$.

_____

_____

PART B: SURDS AND ALGEBRA / 25 marks

15 Complete: $\sqrt{7^2}$ = _____

16 Evaluate $\sqrt{13}$ to 3 significant figures.

_____

17 Complete: $\sqrt{2} \times \sqrt{2}$ = _____

18 Circle the rational numbers:

$\sqrt{25}$ $\sqrt{7}$ $\sqrt{15}$ π $0.1\dot{6}$ 3.51

19 Simplify $3x^2 + 2x - x^2 + 3x$. _____

20 Simplify $3a^2 \times 6a^2$. _____

21 Between which 2 consecutive integers must $\sqrt{76}$ lie? _____

22 Complete: $\sqrt{25 \times 36}$ = _____

23 Expand $3p(r - p)$. _____

24 Find 2 square numbers with a product of:

   a 64 _____

   b 1225 _____

25 In the diagram below, find h in surd form.

26 Expand $(2x - 1)(x + 4)$.

_____

_____

27 Arrange these from smallest to largest:
$\sqrt{98}$, 3.6^2, $\sqrt{81}$, 10.1, $\sqrt{40}$

_____

28 Complete: $(\sqrt{a})^2$ = _____

29 Find a square number and an integer that have a product of:

   a 75 _____

   b 80 _____

30 Is it always true that $\sqrt{a + b} = \sqrt{a} + \sqrt{b}$?

_____

31 Expand $(2k - 5)(2k + 5)$.

_____

32 Is this triangle right-angled?

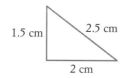

33 a Find x in surd form.

_____

   b Write sin 60° as a fraction. _____

34 Expand $(3x - y)^2$.

_____

35 Is it always true that $\sqrt{a \times b} = \sqrt{a} \times \sqrt{b}$?

_____

36 Find the distance between $(-2, 3)$ and $(5, 6)$ on the number plane (in surd form).

_____

_____

_____

PART C: CHALLENGE   Bonus   / 3 marks

If a sheet of A4 paper is cut in half, the half-sheet is similar to the whole sheet (that is, matching sides are in the same ratio). If the width of the whole sheet is 1 unit, find x, the length of the sheet, as a surd.

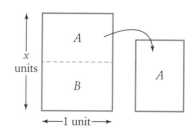

SIMPLIFYING SURDS ①

YOU'LL NEED A RULER TO COMPLETE THIS PUZZLE TO SPELL OUT THE MYSTERY NUMBER.

First complete this pattern of the first 10 square numbers:

1, 4, 9, _____, _____, _____, _____, _____, _____, _____

Simplify the surds in the puzzle below and draw straight lines between the dots to join equivalent surds. Each correct line you draw will go through one letter and one number and these pairs are the solution to the code. Write the numbers corresponding to the letters given in the grid under the puzzle to reveal a famous irrational number.

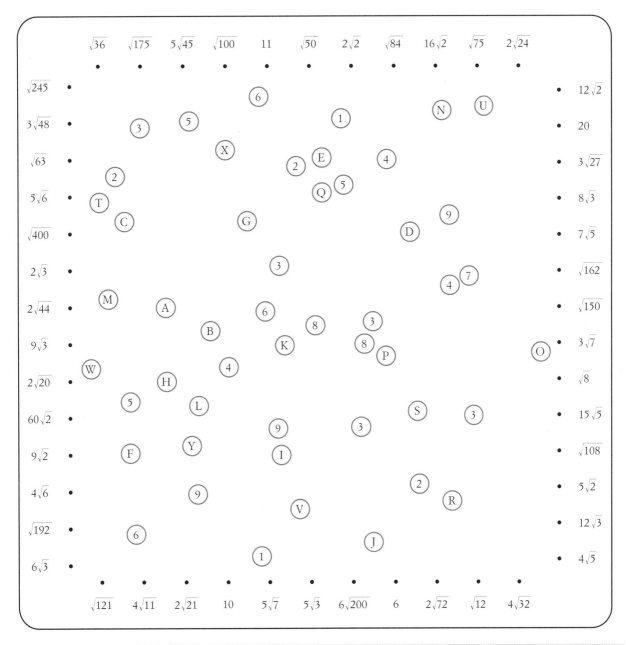

A		B	C	D	E	F	G	H	I	J	K	L	M	N	O	P	Q	R	S	T	U	V	W	X	Y
	•																								

(1) RATIONALISING THE DENOMINATOR

Rationalise the denominator of each surdic expression.

1 $\dfrac{5}{\sqrt{2}}$ _____

2 $\dfrac{3}{\sqrt{6}}$ _____

3 $\dfrac{\sqrt{5}}{\sqrt{3}}$ _____

4 $\dfrac{\sqrt{10}}{2\sqrt{5}}$ _____

5 $\dfrac{2\sqrt{3}}{\sqrt{2}}$ _____

6 $\dfrac{5\sqrt{5}}{2\sqrt{6}}$ _____

7 $\dfrac{2+\sqrt{3}}{\sqrt{5}}$ _____

8 $\dfrac{\sqrt{5}-3}{\sqrt{6}}$ _____

9 $\dfrac{6+\sqrt{12}}{\sqrt{3}}$ _____

10 $\dfrac{7-\sqrt{5}}{3\sqrt{5}}$ _____

11 $\dfrac{4\sqrt{5}-2\sqrt{3}}{\sqrt{2}}$ _____

12 $\dfrac{7\sqrt{10}+3\sqrt{7}}{\sqrt{7}}$ _____

13 $\dfrac{8-\sqrt{6}}{2\sqrt{2}}$ _____

14 $\dfrac{7+3\sqrt{5}}{3\sqrt{10}}$ _____

Extension

Hint: For question **15**, multiply the numerator and denominator by $(5-\sqrt{2})$.

15 $\dfrac{6}{5+\sqrt{2}}$ _____

16 $\dfrac{4}{\sqrt{5}-1}$ _____

17 $\dfrac{\sqrt{3}}{\sqrt{6}+2}$ _____

18 $\dfrac{4\sqrt{6}}{5-\sqrt{5}}$ _____

19 $\dfrac{3\sqrt{2}}{2\sqrt{3}-3}$ _____

20 $\dfrac{6+\sqrt{2}}{4-\sqrt{2}}$ _____

Mixed answers: $\sqrt{70}+3$, $\dfrac{2\sqrt{5}+\sqrt{15}}{5}$, $\dfrac{5\sqrt{30}}{12}$, $\dfrac{5\sqrt{2}}{2}$, $\sqrt{5}+1$, $\dfrac{3\sqrt{2}-2\sqrt{3}}{2}$, $\dfrac{7\sqrt{5}-5}{15}$, $\dfrac{\sqrt{15}}{3}$, $\sqrt{6}$,

$2\sqrt{6}+3\sqrt{2}$, $\dfrac{4\sqrt{2}-\sqrt{3}}{2}$, $\dfrac{5\sqrt{6}+\sqrt{30}}{5}$, $\dfrac{7\sqrt{10}+15\sqrt{2}}{30}$, $2\sqrt{3}+2$, $\dfrac{10-5\sqrt{2}}{3}$, $\dfrac{\sqrt{30}-3\sqrt{6}}{6}$, $\dfrac{\sqrt{2}}{2}$,

$\dfrac{13+5\sqrt{2}}{7}$, $2\sqrt{10}-\sqrt{6}$, $\dfrac{30-6\sqrt{2}}{23}$, $\dfrac{\sqrt{6}}{2}$

9780170454568

THE ANSWERS TO THIS CROSSWORD ARE BELOW.
YOU JUST HAVE TO WORK OUT WHERE THEY GO.
LOOK FOR THE LONG OR SHORT WORDS FIRST.

The answers to this crossword are listed below, in alphabetical order.

Arrange them in the correct places in the puzzle.

ANNULUS	AREA	BASE	CAPACITY
APPROXIMATE	BINOMIAL	DENOMINATOR	DIFFERENCE
EXPAND	EXPRESSION	IRRATIONAL	PRODUCT
QUOTIENT	RATIONAL	RATIONALISE	REAL
ROOT	SIMPLIFY	SQUARE	SURD
UNDEFINED			

(1) SURDS 1

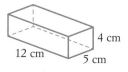

A SURD IS IRRATIONAL, AND ABSURD MEANS 'MAKES NO SENSE'. CAN YOU SEE HOW THESE 2 WORDS ARE RELATED?

Part A	/ 8 marks
Part B	/ 8 marks
Part C	/ 8 marks
Part D	/ 8 marks
Total	/ 32 marks

PART A: MENTAL MATHS

🔢 Calculators not allowed

1

Stem	Leaf
1	3 6
2	1 4 4 7
3	0 2 2 3
4	4 7 8

For the above data, find:

a the mode _____

b the median _____

2 Solve $2m + 8 = 3(2m - 5)$.

_____

_____

3 Find x.

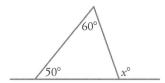

_____

4 What is the centre and radius of the circle with equation $x^2 + y^2 = 36$?

_____

5 Find the surface area of this prism.

4 cm
12 cm
5 cm

_____

_____

_____

6 Find the midpoint of the interval joining points (4, 8) and (–2, 6) on the number plane.

_____

7 Is the age you turn this year discrete or continuous data? _____

PART B: REVIEW

1 Circle the square numbers in this list:

22   49   75   32   100

_____

2 Expand each expression.

a $-6(3p + 4)$

_____

C
S
F

9780170454568

b $(x-3)(x-7)$

_____

_____

_____

3 Simplify each expression.

a $\left(\sqrt{x}\right)^2$ _____

b $\sqrt{5^2}$ _____

4 Circle the surds in this list:

$\sqrt{49}$ $\sqrt{88}$ $\sqrt{81}$ $\sqrt{125}$ $\sqrt{144}$

5 Is each number rational or irrational?

a $\sqrt{99}$ _____

b $\sqrt{289}$ _____

PART C: PRACTICE

> › Symplifying surds
> › Adding and subtracting surds

1 Simplify each expression.

a $\sqrt{160}$

_____

b $\dfrac{\sqrt{250}}{50}$

_____

c $9\sqrt{3} - 4\sqrt{3}$

_____

d $4\sqrt{6} - 5\sqrt{5} + 6\sqrt{6}$

_____

2 (4 marks) Simplify each expression.

a $\sqrt{27} - \sqrt{8} + \sqrt{18}$

_____

_____

_____

b $5\sqrt{20} - 3\sqrt{125}$

_____

_____

_____

PART D: NUMERACY AND LITERACY

1 True or false?

a $\sqrt{x}$ has 2 answers when x is positive

_____

b $\sqrt{x}$ has no value when x is negative

_____

c $\sqrt{x} = x$ for only 2 values of x

_____

d $\sqrt{x}$ is a surd when x is positive

_____

2 Is each number rational (R) or irrational (I)?

a 10π _____

b $0.7\dot{4}$ _____

c $\sqrt[3]{-8}$ _____

d $2 + \sqrt{5}$ _____

Chapter 1 Surds   **7**

1 SURDS 2

I SEE SOME RED QUESTIONS ON THESE PAGES. THIS ALGEBRA NEEDS YOUR CAREFUL BRAIN TRAINING.

Name: _____

Due date: _____

Parent's signature: _____

Part A	/ 8 marks
Part B	/ 8 marks
Part C	/ 8 marks
Part D	/ 8 marks
Total	/ 32 marks

PART A: MENTAL MATHS

🖩 Calculators not allowed

1 (2 marks) Find the area of this triangle.

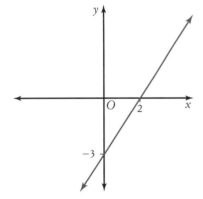

6 cm   10 cm

_____

_____

2 Find the gradient of this line.

_____

3 Simplify 36 : 48 : 18.

_____

4 Expand and simplify $3(1 - 2x) - x(2 - x)$.

_____

_____

_____

5 Factorise $49x^2 - 42x + 9$.

_____

_____

_____

6 (2 marks) 2 coins are thrown together. List all the possible pairings in the sample space (H = heads, T = tails).

_____

_____

PART B: REVIEW

1 Simplify each surd.

a $\sqrt{275}$

_____

b $7\sqrt{48}$

_____

2 (6 marks) Simplify each expression.

a $\sqrt{200} + \sqrt{18}$

_____

_____

b $3\sqrt{5}+\sqrt{50}-2\sqrt{125}$

_____

_____

c $\sqrt{45}-3\sqrt{63}+5\sqrt{80}$

_____

_____

PART C: PRACTICE

 › Multiplying and dividing surds
› Rationalising the denominator

1 Simplify each expression.

a $5\sqrt{10}\times3\sqrt{3}$

_____

b $\dfrac{\sqrt{128}}{\sqrt{2}}$

_____

c $\sqrt{80}\div\left(-4\sqrt{5}\right)$

_____

2 (5 marks) Rationalise the denominator of each surd.

a $\dfrac{5}{\sqrt{2}}$

_____

b $\dfrac{14}{3\sqrt{7}}$

_____

c $\dfrac{5-\sqrt{3}}{2\sqrt{3}}$

_____

_____

PART D: NUMERACY AND LITERACY

1 **a** What is the first step of rationalising the denominator of $\dfrac{20}{\sqrt{5}}$?

_____

_____

b Why is it called 'rationalising the denominator'?

_____

_____

2 Complete each formula.

a $\sqrt{xy}=$ _____

b $\sqrt{\dfrac{x}{y}}=$ _____

c $(a-b)^2=$ _____

3 (3 marks) Expand and simplify each expression.

a $\left(\sqrt{5}-2\right)^2$

_____

_____

b $\left(5\sqrt{2}-4\right)\left(\sqrt{2}+5\right)$

_____

_____

_____

HERE ARE SOME MATHS SKILLS YOU NEED FOR THIS YEAR. PART A IS MIXED SKILLS, PART B IS FOR THE FINANCIAL MATHS TOPIC WE'RE LEARNING.

WS WORKSHEET

PART A: BASIC SKILLS   / 15 marks

1 Calculate, correct to 3 decimal places:

$$\frac{6 \times 9.2}{\sqrt{5 - 2.35}}$$ _____

2 Simplify:

  a $(4x^2)^3$ _____

  b $\left(\frac{2}{3}\right)^3$ _____

3 Expand and simplify:

$2(3d - 5) + 4(d + 4)$

_____

_____

4 What is the formula $V = \pi r^2 h$ used for?

_____

5 Solve $3x^2 = 48$.

_____

6 Simplify:

  a $8 : 28$ _____

  b $\frac{4}{5} : \frac{2}{3}$ _____

7 Which congruence test proves that

  $\triangle ADC \equiv \triangle BCD$ in the rectangle below?

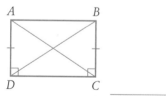

_____

8 Convert 1.5 mL/second to L/hour. _____

_____

9 Find as a surd the distance between (1, 1) and (3, 5) on the number plane. _____

_____

10 Find the area of this trapezium.

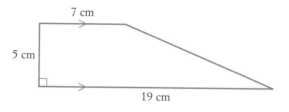

7 cm

5 cm

19 cm

_____

_____

11 Do the diagonals of a parallelogram bisect each other at right angles?

_____

12 Find the value of r in the diagram below.

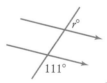

$r°$

111°

_____

13 Evaluate:

$$\frac{4.96 \times 10^7}{3.1 \times 10^{-4}} =$$ _____

PART B: INTEREST /25 marks

14 Complete:

   a $2\frac{1}{2}$ years = _____ months

   b $2\frac{1}{2}$ years = _____ weeks

   c 1 quarter = _____ months

   d 4 years = _____ quarters

   e 9% p.a. = _____% per month

   f 7% p.a. = _____% per quarter

15 Write each percentage as a fraction.

   a 12% _____

   b $12\frac{1}{2}$ % _____

16 Write each percentage as a decimal.

   a 9% _____   **b** $6\frac{1}{2}$ % _____

   c 1.5% _____   **d** 0.8% _____

17 What percentage is $851 of $3700?

_____

18 Evaluate:

   a 8% of $4000 _____

   b 0.05% of $6420 _____

   c $5\frac{3}{4}$ % of $30 000 _____

19 To pay off a loan, Chris pays $138 per fortnight. How much does he pay over $1\frac{1}{2}$ years?

_____

20 Calculate Sandy's total pay if she works 39 hours at $19.20 per hour and 4 hours overtime at time-and-a-half. _____

_____

21 Increase $135 by 7%. _____

_____

22 Calculate the simple interest earned if $3000 is invested at 4% p.a. for 5 years.

_____

23 Toni needs to repay a loan of $2076 over 2 years. Calculate how much she should pay each month.

_____

24 Calculate, correct to the nearest cent:

   a $950 \times (1.12)^7$ _____

   b $5300 \times (0.85)^3$ _____

25 Each week, Ivan pays $264.60 in tax. If this is 35% of his income, then calculate his income.

_____

_____

26 Decrease $2700 by 13%. _____

_____

PART C: CHALLENGE Bonus /3 marks

When Simon was born, his grandparents deposited $5000 into a trust fund for him. Due to interest, the amount in the fund increases by 6% each year. After how many years will the $5000 double in value?

② COMPOUND INTEREST

WHEN INTEREST IS COMPOUNDED, YOU GET INTEREST ON YOUR INTEREST.

Write all money answers correct to the nearest cent. Use 1 year = 52 weeks = 365 days.

1 How many months in:

a 4 years? _____   **b** 7 years? _____

c $3\frac{1}{2}$ years? _____   **d** $1\frac{1}{4}$ years? _____

2 Write as a decimal:

a 4% _____   **b** 6.5% _____

c 13.25% _____   **d** 8.03% _____

e $7\frac{1}{2}$% _____   **f** $3\frac{3}{4}$% _____

g 0.6% _____   **h** 0.148% _____

3 Convert each rate to a monthly interest rate expressed as a decimal.

a 13.5% p.a. _____

b 18% p.a. _____

c 3.15% p.a. _____

d 11.07% p.a. _____

4 Convert each rate to a daily interest rate expressed as a decimal correct to 4 significant figures.

a 15% p.a. _____

b 7.5% p.a. _____

c 21.6% p.a. _____

d 16.72% p.a. _____

5 Convert each rate to a quarterly interest rate expressed a decimal.

a 18% p.a. _____

b 7% p.a. _____

c 10.25% p.a. _____

d 15.46% p.a. _____

6 | **Compound interest formula** |
$A = P(1 + r)^n$

$5000 is invested at 8% p.a. compounded yearly.

Find the final amount at the end of:

a 2 years _____

b 4 years _____

c 5 years _____

7 Calculate the final amount of each investment accumulating compound interest.

a $3500 invested at 7% p.a. for 5 years

_____

b $9900 invested at 10.2% p.a. for 7 years

_____

c $25 000 invested at 8.95% for 3 years

_____

d $12 000 invested at 4% p.a. compounded half-yearly for 2 years

_____

e $15 500 invested at 9.6% p.a. compounded monthly for 4 years

_____

f $37 000 invested at 11.7% p.a. compounded quarterly for 3 years

_____

8 Calculate the compound interest earned when:

a $4000 is invested at 6% p.a. for 4 years

_____

b $21 500 is invested at 8.5% p.a. compounded half-yearly for 6 years

_____

c $32 000 is invested at 9.2% p.a. compounded quarterly for 2 years

_____

d $9750 is invested at 7.05% p.a. compounded monthly for $1\frac{1}{2}$ years

_____

9 Calculate the principal to be invested at compound interest to reach each final amount.

a $3200 in 3 years invested at 9% p.a.

_____

b $9500 in 4 years invested at 7.3% p.a.

_____

c $15 000 in 2 years invested at 8% p.a.

_____

10 Find the values missing from this table.

	Principal	Rate (% p.a.)	Time	Compounded	Final amount	Interest
a	$5500	7%	4 years	Yearly		
b		6.4%	6 years	Half-yearly	$9300	
c	$20 000	12.6%	3.5 years	Monthly		
d		9%	2 years	Monthly	$25 600	
e		14.8%	$2\frac{3}{4}$ years	Quarterly	$12 220	

Chapter 2 Interest and depreciation **13**

② DEPRECIATION

WS WORKSHEET

DEPRECIATION IS THE OPPOSITE OF COMPOUND INTEREST, WHEN THE VALUE OF AN ITEM DECREASES BY THE SAME PERCENTAGE.

1 The purchase price of a boat is $48 000. If the boat depreciates by 10% p.a., calculate its value after 5 years.

_____

2 A $34 000 new car depreciates by 11% p.a.

a Calculate the value of the car after 8 years.

_____

b Calculate the amount by which the vehicle depreciates in 8 years.

_____

3 A local business has purchased office furniture to the value of $26 500.
Find the furniture's value after 3 years of use if it depreciates at 28.5% per annum.

_____

4 A factory depreciates in value by 4.5% per annum. If its current value is $315 000, find its value after 10 years.

_____

5 A construction company purchases a new grader for $356 400. If the grader depreciates at a rate of 22% p.a.:

a calculate its value after 1 year

_____

b calculate its value after the second year

_____

c by how much has the grader depreciated in the second year?

_____

6 Thomas buys a new snowmobile for $11 820. It depreciates at a rate of 15.5% per annum.

a Find the value of Thomas' snowmobile after 2 years.

_____

b Find the value of Thomas' snowmobile after 5 years as a percentage of its original value, correct to one decimal place.

_____

7 A Blu-ray DVD player, originally costing $480, depreciates at 10% per annum. What percentage of its original value is its value after:

a 1 year? _____

b 2 years? _____

c 5 years? _____

8 Patrick spent $18 500 on equipment for his gardening business. The equipment depreciates at 18.5% per year.

a Find the value of the equipment after 4 years.

_____

b Find the amount of depreciation of the equipment after 2 years.

_____

c Find how long it will take Patrick's equipment to have a value less than $5000.

_____

9 A company car purchased for $46 000 depreciates at 11.4% per annum.

a Calculate the value of the car after 3 years.

_____

b Calculate the total depreciation over the first 7 years.

_____

c How long will it take for the car to reach a value below $10 000?

_____

_____

10 A front-end loader was purchased for $415 300. It depreciates at 20% per annum. The owner sells the loader for $245 000 after 3 years. Was there a profit or loss made on the sale of the front-end loader?

_____

_____

_____

11 The Australian Taxation Office allows depreciation on tools of trade as a legitimate tax deduction. The depreciation rate is 13.5% per annum. A bricklayer purchases tools to the value of $18 300. When the value falls below $6000, the bricklayer is allowed to write off the tools on the next year's tax return. Calculate when the tools can be written off.

_____

_____

_____

_____

12 Adrian bought a new car for $27 900. The salesperson claimed that, at 10% p.a. depreciation, the car would decrease in value by $2790 per year. Do you agree or disagree with this statement? Give reasons.

_____

_____

_____

_____

UNSCRAMBLE THE WORDS FROM THIS TOPIC NEXT PAGE TO COMPLETE THIS PUZZLE.

Clues across

1 PETMARYEN

4 TENSTENVIM

6 SORGS

7 AGEW

10 NIPLIPCAR

11 DAGLION

12 TRYGOLFHINT

16 TEAR

18 LAANNU

20 METR

22 CUPNODOM

23 LAFIN

25 VEALE

26 ALYSRA

Clues down

2 IMPELS

3 TALF

5 ROVEITEM

8 ATX

9 TOPSIDE

13 MINICOSMOS

14 ICEDDONUT

15 ENT

17 NOMICE

19 NITTREES

21 TNCREEP

23 LAFORUM

24 MYLONTH

27 GAPY

28 LUERYTQAR

29 LACANBE

30 WOKRECIPE

② COMPOUND INTEREST

THIS IS YOUR WEEKLY HOMEWORK ASSIGNMENT, COVERING THE CURRENT TOPIC AS WELL AS MIXED REVISION. NO CALCULATORS IN PART A!

Name:

Due date:

Parent's signature:

Part A	/ 8 marks
Part B	/ 8 marks
Part C	/ 8 marks
Part D	/ 8 marks
Total	/ 32 marks

HW HOMEWORK

PART A: MENTAL MATHS

🧮 Calculators not allowed

1 Find 5% of $4000. _____

2 Between which 2 consecutive whole numbers is the value of $\sqrt{70}$?

_____

3 Calculate, as a surd, the distance between the points $(-3, -2)$ and $(-2, 7)$ on the number plane.

_____

4 Write the formula for a closed cylinder's:

a surface area _____

b volume _____

5 For the data in the table, find:

x	y
2	3
3	8
4	6
5	6
Total	23

a the median _____

b the mode _____

c the range _____

PART B: REVIEW

1 Convert each percentage to a decimal.

a 17.5% _____

b 5% _____

2 Increase $4500 by 2.5% _____

3 Complete:

a 1 year = _____ weeks

b 1 year = _____ months

c 1 year = _____ days

4 Evaluate correct to the nearest cent:

a $50\,000 \times (1.05)^5$ _____

b $34\,300 \times (1.071)^{10}$ _____

9780170454568

PART C: *PRACTICE*

> › Simple interest
> › Compound interest

1 Calculate the simple interest earned on $25 000 invested for 5 months at 5.3% p.a.

_____

2 (2 marks) The simple interest on a loan of $15 960 over 5 years is $5200.

Calculate the interest rate p.a., correct to one decimal place.

_____

3 Calculate the value of an investment of $18 250 at 3.2% p.a. interest compounded annually after 3 years.

_____

4 For an investment of $20 000 for 4 years compounded monthly at 4.5% p.a., find:

a the total amount

_____

b the interest earned

_____

5 (2 marks) $7400 is invested for 2 years at 3.8% p.a., compounded quarterly. Calculate the interest earned.

_____

PART D: *NUMERACY AND LITERACY*

1 a What type of interest is calculated only on the original principal?

_____

b What type of interest is used in the formula $A = P(1 + r)^n$?

_____

c In the formula $I = Prn$, what does the n stand for?

_____

d In the formula $A = P(1 + r)^n$, what does the P stand for?

_____

2 $38 000 is invested for 6 months at 17.5% p.a.

_____

a Calculate the simple interest.

_____

b Calculate the total amount.

_____

3 For an investment of $16 500 for one year at 1.6% p.a. compounded daily, calculate:

a the total amount

_____

b the compound interest

_____

9780170454568

HI, I'M MS LEE. THIS ASSIGNMENT PREPARES YOU FOR THE 'GRAPHING LINES' TOPIC.

PART A: BASIC SKILLS   / 15 marks

1 Write $\dfrac{8}{11}$ as a decimal. _____

2 Find the simple interest earned on $5000 invested for $3\dfrac{1}{2}$ years at 6.2% p.a.

_____

3 Complete: 1 m³ can hold _____ L.

4 Is $\sqrt{2.5}$ rational or irrational?

_____

5 Evaluate $\dfrac{9}{2\sqrt{3}}$, correct to 3 significant figures.

_____

6 How much petrol can I buy for $50 at 142.3 cents/litre? Answer to the nearest 0.1 L.

_____

7 Write the formula for the surface area of a cylinder.

_____

8 a Write 0.041 05 in scientific notation.

_____

b How many significant figures has 0.041 05?

_____

9 Expand $3(3y + 1)(3y - 1)$. _____

_____

10 What is the angle sum of a quadrilateral?

_____

_____

11 Expand $(5u - 4)^2$. _____

12 Simplify $\dfrac{3m^2}{4} \div \dfrac{15m}{8}$. _____

13 Solve $4(d + 5) = 32$. _____

_____

14 Find the value of x if these triangles are similar.

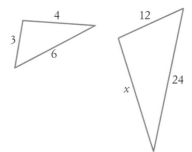

PART B: COORDINATE GEOMETRY

/ 25 marks

15 Complete the table below for $y = 4x - 5$.

x	-2	-1	0	1
y				

16 Evaluate $\dfrac{d - b}{c - a}$ if $a = -7, b = 2, c = -1$ and $d = 0$.

_____

Questions **17** to **19** refer to this diagram.

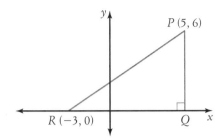

17 If △PQR is right-angled, write the coordinates of Q. _____

18 Find the area of △PQR. _____

19 For the hypotenuse RP, find its:

   a length _____

   _____

   b midpoint _____

   _____

   c gradient. _____

   _____

20 Graph $y = -2$ and $y = -3x$ on a number plane.

21 Write one property of the sides of a:

   a trapezium

   _____

   _____

   b rhombus

   _____

   _____

   c parallelogram.

   _____

   _____

22 Find the length of the hypotenuse of a right-angled triangle as a surd if the other sides are 3 cm and 6 cm. _____

   _____

23 Write the equation of the horizontal line that goes through $(0, -4)$.

   _____

24 Draw a line whose gradient is negative.

WORKSHEET

WS

25 Graph $y = 2x - 3$ on a number plane and find its gradient and y-intercept.

_____

26 Find y when $x = 2$ if:

a $y = 8 - 3x$ _____

b $y = \dfrac{x}{4} + 5$ _____

c $y = \dfrac{1}{2}x^2$ _____

d $y = x^3 - 4$ _____

27 Does the point $(4, 7)$ lie on the line

$y = -2x + 15$? _____

_____

28 Show that $y = \dfrac{1}{2}x + 3$ can be rewritten as

$2x - 4y + 12 = 0$.

_____

_____

_____

_____

_____

PART C: CHALLENGE  Bonus / 3 marks

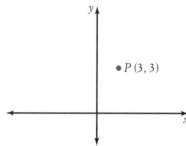

- How many points on the number plane (with integer coordinates) are exactly 5 units from $P(3, 3)$?

  _____

- If *all* points that are 5 units from P are graphed (including those with non-integer coordinates), what shape would result?

  _____

WRITING EQUATIONS OF LINES ③

FIND THE EQUATION OF ANY LINE
THAT FITS EACH DESCRIPTION,
IN THE FORM $y = mx + c$.

1 $(1, 3)$ lies on the line.	**2** Parallel to $y = 6x - 7$	**3** Parallel to the x-axis.	**4** Has a negative gradient and passes through $(0, 0)$.
5 Perpendicular to $y = -3x - 1$.	**6** The y-intercept is 5.	**7** The gradient is $\dfrac{1}{6}$.	**8** Perpendicular to $3x - 2y + 4 = 0$.
9 Passes through $(-1, 4)$ and is parallel to the y-axis.	**10** Steeper than $y = 2x - 4$.	**11** The x-intercept is -7.	**12** Passes through $\left(\dfrac{1}{2}, 5\right)$.
13 Parallel to $2x - y + 7 = 0$.	**14** Perpendicular to the x-axis.	**15** Has the same y-intercept as $x - 3y + 15 = 0$.	**16** Perpendicular to the y-axis and passes through $(-3, -5)$.
17 Has a positive gradient and passes through $(0, -5)$.	**18** Parallel to $y = -5x + 7$ and cuts the x-axis at 2.	**19** The x-intercept is 10 and the y-intercept is -3.	**20** Steeper than $y = -\dfrac{x}{2} - 4$.
21 $(2, -1)$ lies on the line and it has a gradient of 3.	**22** Perpendicular to $y = \dfrac{2x}{3} + 7$ with a negative y-intercept.	**23** Passes through $(0, -4)$ and $(9, 0)$.	**24** Passes through $(1, -5)$ and $(5, -9)$.

③ LINEAR EQUATIONS CODE PUZZLE

CAN YOU DECODE THIS MESSAGE?
MATCH EACH EQUATION TO ITS
GRAPH OR FEATURE NEXT PAGE.

6	12	7	10

14	15	14

2	17	11

25	7	9	7	8	8	11	8

8	15	17	11

4	7	1

10	2

7	17	2	10	12	11	9

7	10

10	12	11

				?
25	7	9	10	1

'
11	13	20	18	4	11

	,
21	11

1	2	18

8	2	2	24

23	11	9	1

5	7	21	15	8	15	7	9

3	18	10

	,
6	11

23	11

17	11	23	11	9

		,
21	11	10

12	7	23	11

	?'
6	11

The numbers in the grid above match the question numbers. Write each of the following linear equations in the form $y = mx + c$ (where appropriate) and match each one with its correct feature or graph on the next page. Each answer has a corresponding letter. Fill in the grid above with the letters that match the questions, to decode a riddle.

1 $2x + y - 1 = 0$

_____

2 $x + y + 1 = 0$

_____

3 $x + 2y - 1 = 0$

_____

4 $y = 7$

_____

5 $x = 1$

_____

6 $y = x$

_____

7 $x + 5y = 0$

_____

8 $y = x - 3$

_____

9 $4x - 2y + 6 = 0$

_____

10 $y = \frac{1}{2}x + 1$

_____

11 $y = -2$

_____

12 $2x - 3y + 3 = 0$

_____

13 $2x - y - 2 = 0$

_____

14 $y = -3x$

_____

15 $x = -2$

_____

16 $y = 3x - 4$

_____

17 $x - y + 4 = 0$ _____

18 $8x + 2y - 20 = 0$ _____

19 $2x - 3y - 9 = 0$ _____

20 $2x - y - 1 = 0$ _____

21 $x + y - 2 = 0$ _____

22 $y = x + \dfrac{1}{2}$ _____

23 $3x - 2y - 14 = 0$ _____

24 $5x + y - 2 = 0$ _____

25 $y = 0$ _____

T

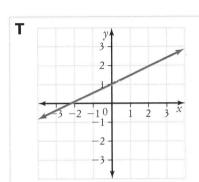

Y

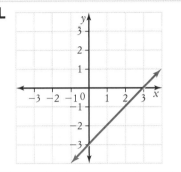

L

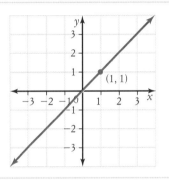

F

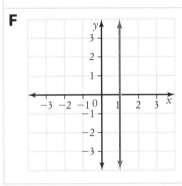

X

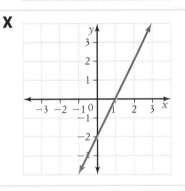

W

C $m = 2, c = -1$	**K** $m = -5$	**G** $m = 1, c = \dfrac{1}{2}$	**D** $m = -3$
A $m = -\dfrac{1}{5}$	**H** $m = \dfrac{2}{3}, c = 1$	**V** $m = \dfrac{3}{2}$	**O** $m = c = -1$
N $c = 4$	**S** $c = 7$	**J** $c = -4$	**U** $c = 10$
I vertical line passing through $(-2, 0)$		**M** $y = 2 - x$	**R** $y = 2x + 3$
B $y = -\dfrac{1}{2}x + \dfrac{1}{2}$	**Z** $y = \dfrac{2}{3}x - 3$	**E** $y = -2$	**P** the x-axis

③ COORDINATE GEOMETRY

AS IN SPORT, DRILL AND PRACTICE ARE IMPORTANT IN MATHS. PART B OF THIS HOMEWORK ASSIGNMENT IS REVIEW OF PREVIOUS WORK, PART C IS PRACTICE OF THIS WEEK'S WORK.

Name:

Due date:

Parent's signature:

Part A	/ 8 marks
Part B	/ 8 marks
Part C	/ 8 marks
Part D	/ 8 marks
Total	/ 32 marks

PART A: MENTAL MATHS

Calculators not allowed

1 A TV costs $400 after 20% discount. What was its original price?

_____

_____

2 A fridge has its price reduced from $800 to $680. What is the percentage discount?

_____

3 Convert:

a 60 km/h = _____ m/min

b 4 mL/min = _____ L/h

4 (4 marks) Find the value of each variable, giving reasons.

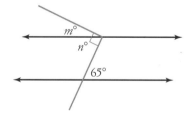

PART B: REVIEW

1 (4 marks) Complete each table of values.

a $y = x - 2$

x	-2	-1	0	1
y				

b $y = 4x - 3$

x	-1	0	1	2
y				

2 State whether each line's gradient is positive, negative or neither.

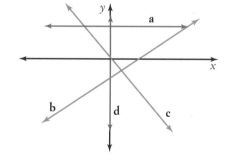

a _____

b _____

c _____

d _____

9780170454568

PART C: PRACTICE

📝 › Length, midpoint, gradient
› Graphing $y = mx + c$

1

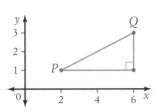

a Find the length of PQ as a surd.

_____

_____

b Find the midpoint of PQ.

_____

c Find the gradient of PQ.

_____

2 (3 marks) Graph $y = -3x + 6$ on the number plane and write its x-intercept and y-intercept.

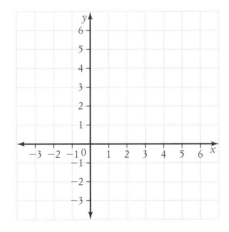

_____

_____

_____

_____

3 What is the gradient of the line with equation $y = -3x + 6$?

_____

4 Test whether $(5, -9)$ lies on the line with equation $y = -3x + 6$.

_____

_____

PART D: NUMERACY AND LITERACY

1 a Draw a quadrilateral $ABCD$ with vertices at $A(0, 4)$, $B(6, 1)$, $C(2, -3)$, $D(-2, -1)$.

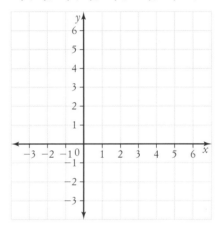

b Find the gradient of the sides of $ABCD$.

$m_{AB} =$ _____

$m_{CD} =$ _____

$m_{AD} =$ _____

$m_{BC} =$ _____

c What type of quadrilateral is $ABCD$?

_____

2 Write the equation of a line with gradient -1 and y-intercept -2.

_____

3 Write the equation of the vertical line that goes through $(3, -4)$.

_____

HOMEWORK

③ GRAPHING LINES

PART D QUESTIONS CAN BE COMPLEX BECAUSE THEY ASK YOU TO WRITE ABOUT YOUR MATHS, USING THE RIGHT TERMINOLOGY.

Name:

Due date:

Parent's signature:

Part A	/ 8 marks
Part B	/ 8 marks
Part C	/ 8 marks
Part D	/ 8 marks
Total	/ 32 marks

PART A: *MENTAL MATHS*

🚫 Calculators not allowed

1 Factorise each expression.

a $49m^2 - 35m$

_____

b $8(x + 2) - p(2 + x)$

_____

2 Simplify each expression.

a $\dfrac{56x^8y^6}{\left(2x^2y^2\right)^2}$

_____

b $\left(\dfrac{3}{4m^2}\right)^{-2}$

_____

3 For this stem-and-leaf plot, identify any:

Stem	Leaf
1	2 5 6
2	2 3 6 7 8
3	0 4 4 8 9 9 9
4	1 2 4 5 5 6 7 8 8 9
5	0 1
6	9

a outliers _____

b clusters _____

4 (2 marks) If $P = 50$ and $L = 4$, find W if $P = 2(L + W)$.

_____

_____

_____

PART B: *REVIEW*

1 a Find the gradient of the line with equation $y = 2x - 6$.

_____

b What is the gradient of a line that is perpendicular to $y = 2x - 6$?

_____

2 Write the equation of a line with gradient $\dfrac{1}{2}$ and y-intercept 7.

_____

9780170454568

3 Graph the line $y = -2$ on the number plane.

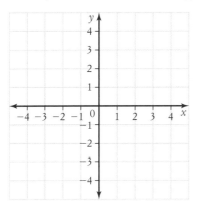

4 (4 marks) Find the gradients of each pair of lines, and then determine whether they are parallel or perpendicular, giving reasons.

a $y = 4x + 13$ $m_1 =$ _____

   $y = 4x - 3$ $m_2 =$ _____

_____

b $2x + 4y - 16 = 0$ $m_1 =$ _____

   $-6x + 3y + 3 = 0$ $m_2 =$ _____

_____

_____

PART C: PRACTICE

> Parallel and perpendicular lines
> Finding the equation of a line

1 Find the equation of this line.

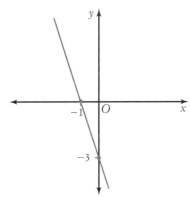

2 (3 marks) Find the equation of the line that is parallel to:

a $y = 5x - 2$ and has a y-intercept of 3

_____

_____

b $2x - 2y - 8 = 0$ and passes through (2, 1)

_____

_____

3 (4 marks) Find the equation of the line that is perpendicular to:

a $y = 5x - 2$ and has a y-intercept of 3.

_____

_____

b $y = -3x - 4$ and passes through the midpoint of (6, –8) and (0, –4)

_____

_____

PART D: NUMERACY AND LITERACY

1 Complete: If 2 lines have gradients m_1 and m_2 and

a $m_1 = m_2$, then they are

_____

b $m_2 = -\dfrac{1}{m_1}$, then they are

_____

A (3, 2)

B (−1, −4)

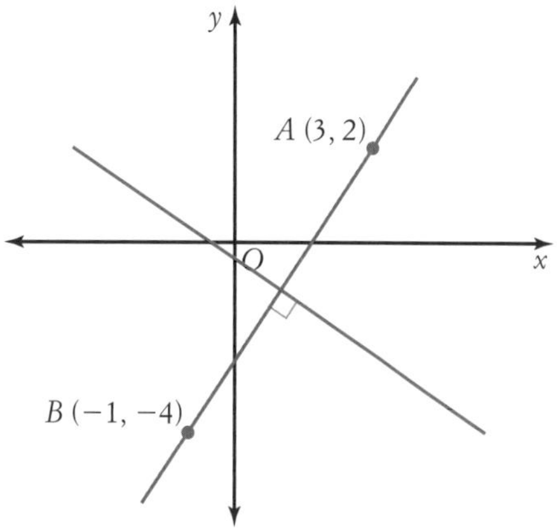

a Find the gradient of interval *AB*.

_____

b Find the midpoint of *AB*.

_____

c The red line is perpendicular to *AB* and passes through its midpoint. What is its gradient?

_____

d Find the equation of the red line.

_____

_____

_____

(−2, 1)

g

k

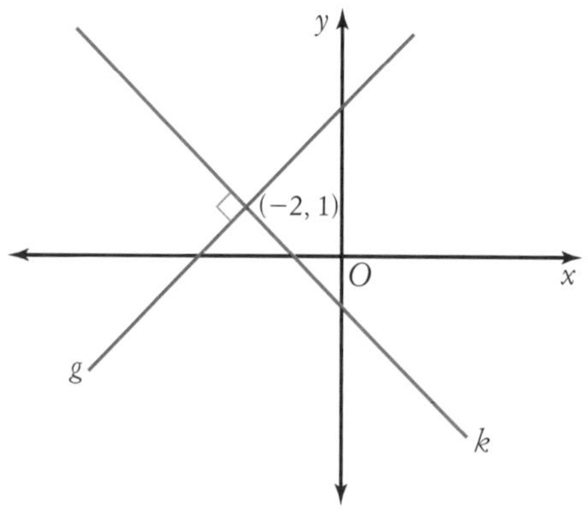

a Find the equation of line *g* if its gradient is $\frac{1}{2}$.

_____

_____

_____

b Find the equation of line *k*.

_____

_____

_____

9780170454568

THOSE ARE THE WORDS, HERE IS THE PUZZLE. YOU KNOW WHAT TO DO.

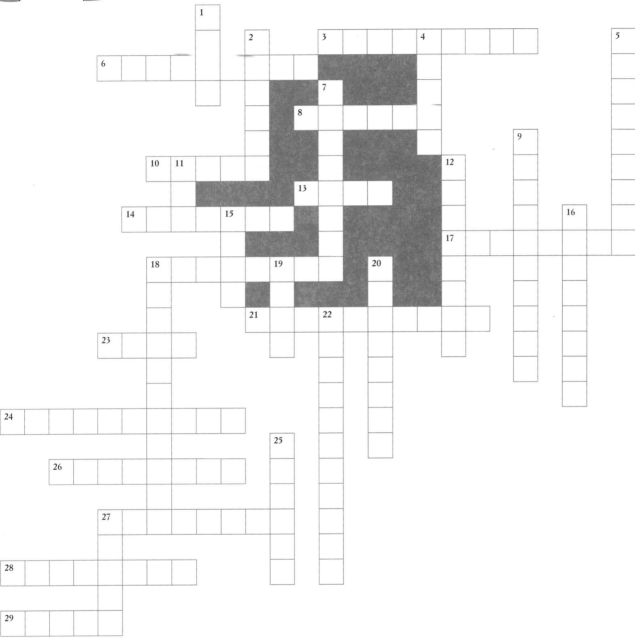

AXES
CARTESIAN
CONSTANT
COORDINATES
EQUATION
EXACT
FORM
GENERAL
GRADIENT
GRAPH
HYPOTENUSE

INCLINATION
INTERCEPT
INTERVAL
LENGTH
LINE
LINEAR
MIDPOINT
NEGATIVE
ORIGIN
PARALLEL
PLANE

POINT
POSITIVE
PYTHAGORAS
RECIPROCAL
RISE
RUN
STEEPNESS
SURD
VERTICAL

(4) STARTUP ASSIGNMENT 4

THIS ASSIGNMENT CAN BE DONE AT THE START OF THE TOPIC BECAUSE IT REVISES AREA AND VOLUME SKILLS YOU'LL NEED FOR THAT TOPIC.

PART A: BASIC SKILLS

/ 15 marks

1 Draw an obtuse-angled isosceles triangle.

2 Simplify $(5k)^{-2}$. _____

3 For the values 19, 12, 11, 17, 11, 12, 13, 11, find:

  a the mean _____

  b the median _____

  _____

4 What is the highest common factor of $18a^2b$ and $12abc$? _____

5 Find the size of the angle x in the diagram below. _____

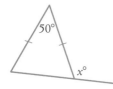

6 Write $\dfrac{17}{20}$ as a percentage. _____

7 Write 37 800 000 in scientific notation.

  _____

8 Solve $3x - 5 = x + 9$. _____

  _____

  _____

9 For the points $A(1, 7)$ and $B(-3, 15)$, find:

  a the length of AB, to 2 decimal places

  _____

  _____

  _____

  b the gradient of AB. _____

10 Simplify $\dfrac{10ab}{2a^2}$. _____

11 Convert $\dfrac{5}{6}$ to a decimal. _____

12 Jane earned \$686.70 for selling \$9810 worth of cosmetics. What was her percentage commission? _____

13 Find d if these triangles are similar.

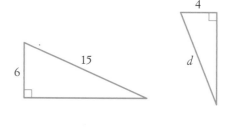

  _____

  _____

9780170454568

PART B: AREA AND VOLUME

/ 25 marks

Round answers to 2 decimal places where necessary.

14 a Draw a square pyramid.

b Draw its net.

15 A circle has a radius of 6 cm. Find, correct to 2 decimal places, its:

a circumference _____

b area _____

16 A cube has sides of length 8 cm. Find its:

a surface area _____

b volume _____

17 a How many cm² in 1 m²? _____

b How many litres in 1 m³? _____

18 a Name this solid. _____

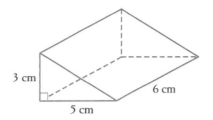

b How many faces does it have? _____

c What shapes are its faces?

_____

d Find the volume of the solid. _____

19 If $r = 3$ and $h = 9$, evaluate correct to 2 decimal places:

a $\dfrac{1}{3}\pi r^2 h$ _____

b $\dfrac{4}{3}\pi r^3$ _____

20 Write the formula for the area of a trapezium.

_____

21 a Name this shape.

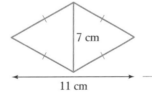

b Find the shape's area. _____

22 Find the perimeter of this triangle.

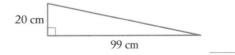

23 Find the base length of a parallelogram with a perpendicular height of 4 cm if its area is 28 cm². _____

24 Write one difference between a prism and a pyramid.

_____

_____

_____

_____

25 Find the radius of this cone.

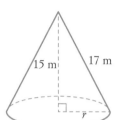

15 m   17 m

r

_____

26 Calculate each area below.

a

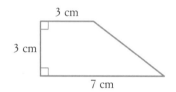

3 cm

3 cm

7 cm

_____

_____

_____

b

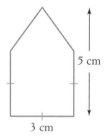

5 cm

3 cm

_____

_____

_____

c

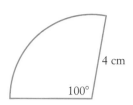

4 cm

100°

_____

_____

_____

27 Find the length of a rectangle if its length is double its width and its area is 18 cm².

_____

_____

_____

_____

PART C: CHALLENGE   Bonus   / 3 marks

A cylinder holds 3 tennis balls neatly.

What fraction of the cylinder's volume is taken up by the tennis balls if the volume of a sphere with radius r is $V = \dfrac{4}{3}\pi r^3$?

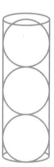

9780170454568

Teacher's tickbox

For each shape:

❏ find the volume   ❏ find the surface area

1

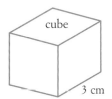

cube

3 cm

2

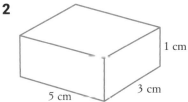

1 cm

5 cm   3 cm

3

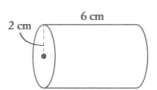

6 cm

2 cm

4

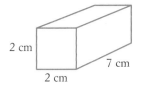

2 cm

2 cm   7 cm

5

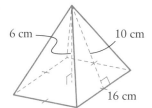

6 cm   10 cm

16 cm

6

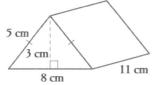

5 cm

3 cm

8 cm   11 cm

7

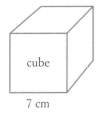

cube

7 cm

8

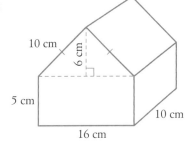

10 cm

6 cm

5 cm

16 cm   10 cm

9

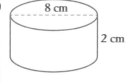

8 cm

2 cm

10

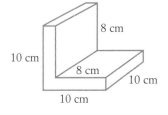

8 cm

10 cm

8 cm   10 cm

10 cm

11

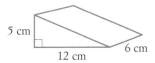

5 cm

12 cm   6 cm

12

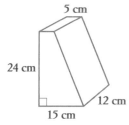

5 cm

24 cm

15 cm   12 cm

13

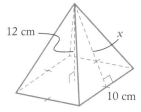

12 cm   x

10 cm

14

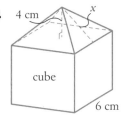

4 cm   x

cube

6 cm

15

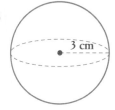

3 cm

16

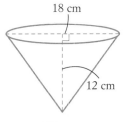

18 cm

12 cm

17

2 cm

18

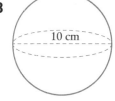

10 cm

19

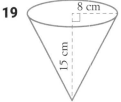

8 cm

15 cm

20

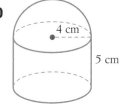

4 cm

5 cm

44 BACK-TO-FRONT PROBLEMS (ADVANCED)

THEY'RE BACK-TO-FRONT BECAUSE ONLY THE ANSWERS ARE GIVEN. YOU HAVE TO FIND SUITABLE MEASUREMENTS FOR EACH SOLID.

1

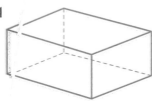

Volume = 160 cm^3

2

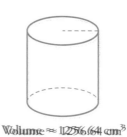

Volume ≈ 1256.64 cm^3

3

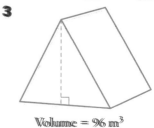

Volume = 96 m^3

4

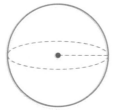

Volume ≈ 1436.76 cm^3

5

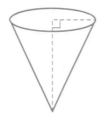

Volume ≈ 183.26 cm^3

6

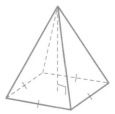

Volume = 588 cm^3

7

12 cm   4 cm

Volume = 320 cm^3

8

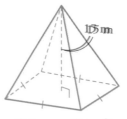

15 m

Volume = 720 m^3

9

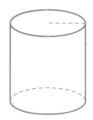

Volume ≈ 311.02 m^3

10

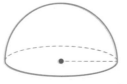

Volume ≈ 1526.81 cm^3

11

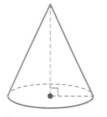

Volume ≈ 117.29 cm^3

12

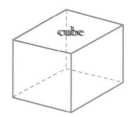

cube

Surface area = 486 cm^3

13

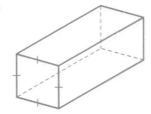

Surface area ≈ 288 cm^3

14

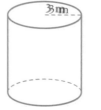

3 m

Surface area ≈ 169.65 m^3

15

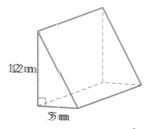

12 m

5 m

Surface area = 660 m^3

HI, ZINA HERE. HERE'S ANOTHER CROSSWORD WHERE THE ANSWERS ARE GIVEN!

The answers to this crossword puzzle are listed below in alphabetical order.

Arrange them in the correct places, leaving out any hyphens.

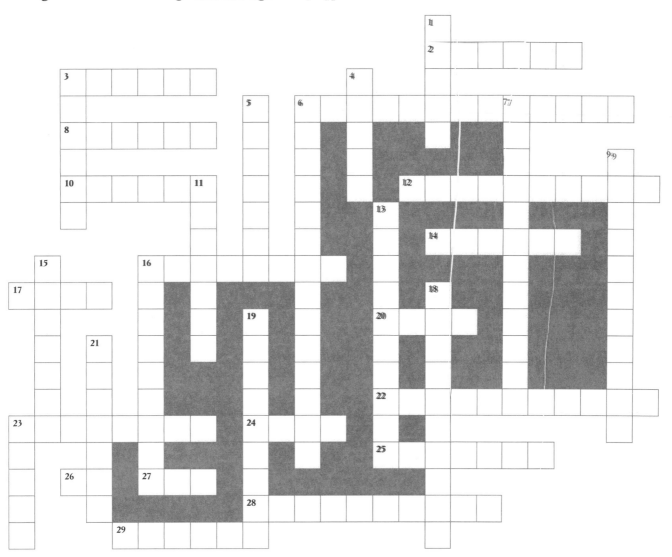

APEX	CAPACITY	COMPOSITE	CLOSED	CONE
CROSS-SECTION	CUBED	DIAMETER	DIMENSION	FORMULA
HEIGHT	HEMISPHERE	LENGTH	NET	OBLIQUE
OPEN	PARALLELEPIPED	PERPENDICULAR	PI	PRISM
PYRAMID	PYTHAGORAS	RADIUS	RECTANGULAR	SEMICIRCLE
SLANT	SPHERE	SQUARE	SQUARED	SURFACE
TRAPEZOIDAL	TRIANGULAR	VOLUME		

④ SURFACE AREA

HEY, I'M MITCH. THIS ASSIGNMENT COVERS MENTAL MATHS, AREA AND SURFACE AREA.

Part A	/ 8 marks
Part B	/ 8 marks
Part C	/ 8 marks
Part D	/ 8 marks
Total	/ 32 marks

PART A: MENTAL MATHS

🚫 Calculators not allowed

1 Solve $\dfrac{x}{7} = \dfrac{x-2}{5}$.

_____

_____

2 At a New Year's sale, bags are discounted by 25%. What is the discount price of a bag marked at $150?

_____

3 For this triangle, write as a fraction:

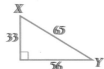

a sin X _____

b cos Y _____

c tan Y _____

4 These dot plots show the ages of students in 2 Taekwondo classes.

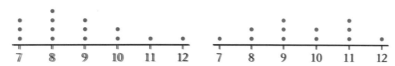

a Which class had more consistent ages?

_____

b Which class had more older students?

_____

5 The scale of a plan is 1 : 2000. What distance in centimetres on the plan would represent a real distance of 50 m?

_____

PART B: REVIEW

1 Find the area of each shape (correct to 2 decimal places for **c** and **e**).

a

_____

_____

b

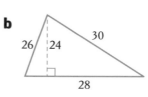

_____

_____

C
S
F

9780170454568

c

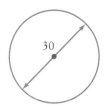

30

_____

_____

d

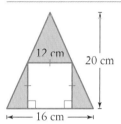

12 cm

20 cm

16 cm

_____

_____

e (2 marks)

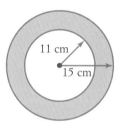

11 cm

15 cm

_____

_____

2 Find the value of each variable.

a

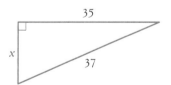

35

x

37

_____

_____

b

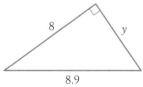

8

y

8.9

_____

_____

PART B: **PRACTICE**

📝 › Surface areas of prisms, cylinders, composite solids

1 (4 marks) Calculate the surface area of each solid (correct to 3 significant figures for **b**). All measurements are in centimetres.

a

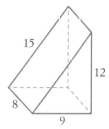

15

12

8

9

_____

_____

_____

b

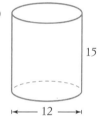

15

12

_____

_____

_____

2 (4 marks) The depth of water in this swimming pool ranges from 1 m to 3 m. Calculate, correct to 2 decimal places:

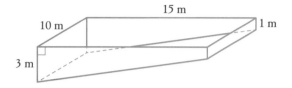

a the area of its slanted floor

_____

_____

_____

_____

b its total surface area

_____

_____

_____

4 (4 marks) Calculate the surface area of each solid, correct to one decimal place. All measurements are in metres.

a

_____

_____

_____

b half-cylinder with open top, one end open

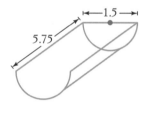

_____

_____

_____

PART D: NUMERACY AND LITERACY

1 What is the **surface area** of a solid?

_____

_____

2 (2 marks) Complete: A prism has the same _____ along its length, and each one is a _____

3 The surface area of a cylinder has the formula:

$SA = 2\pi r^2 +$ _____

Name:

Due date:

Parent's signature:

Part A	/ 8 marks
Part B	/ 8 marks
Part C	/ 8 marks
Part D	/ 8 marks
Total	/ 32 marks

THIS 'ADVANCED' ASSIGNMENT COVERS THE SURFACE AREA OF A PYRAMID, CONE AND SPHERE.

PART A: MENTAL MATHS

🚫 Calculators not allowed

1 In a school of 819 students, girls and boys are in the ratio 4 : 3. How many girls are there?

_____

2 Simplify each expression.

a $(-2a)^{-4}$

_____

b $\left(49r^2\right)^{\frac{3}{2}}$

_____

3 Simplify $\sqrt{48} - \sqrt{12}$.

_____

4 For the interval joining the points $(-3, 7)$ and $(9, 11)$ on the number plane, find its:

a midpoint _____

b length as a surd _____

_____

c gradient _____

5 2 dice are rolled together. Find the probability of rolling a total of 10.

_____

9780170454568

PART B: REVIEW

1 (4 marks) Find the surface area of each prism.

a

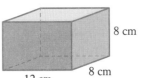

12 cm   8 cm   8 cm

_____

_____

_____

_____

b

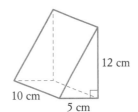

10 cm   5 cm   12 cm

_____

_____

_____

_____

2 (2 marks) Find, correct to one decimal place, the surface area of this half-cylinder.

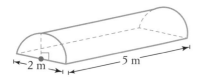

_____

_____

_____

_____

3 (2 marks) Find the area of this sector, correct to the nearest square metre.

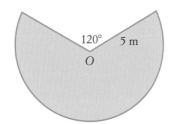

_____

_____

2

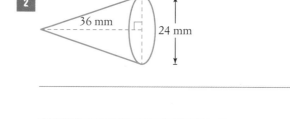

_____

_____

_____

_____

3

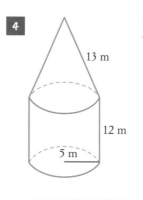

_____

_____

_____

_____

4

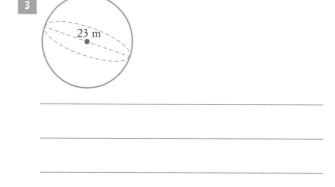

PART C: PRACTICE

📝 › Surface areas of pyramids, cones, spheres, composite solids

Find the surface area of each solid, correct to the nearest square unit.

1

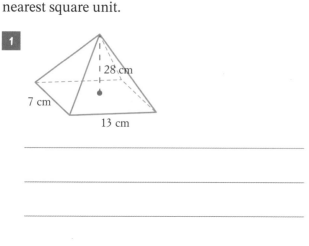

_____

_____

_____

_____

_____

_____

_____

_____

_____

9780170454568

HW HOMEWORK

PART D: NUMERACY AND LITERACY

1 a Which solid has a polygon for its base and triangular faces that meet at a point?

_____

b What is the name of that point? _____

2 How is the slant height of a cone different from its perpendicular height?

_____

_____

_____

3 Explain why the surface area of a hemisphere is $3\pi r^2$, where r is its radius.

_____

_____

_____

4 Find, correct to 2 decimal places, the surface area of each solid.

a

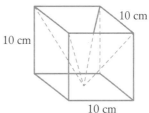

_____

_____

_____

_____

b

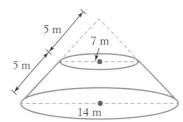

_____

_____

_____

_____

④ VOLUME 2

THE CORRECT WAY TO SAY M³ IS 'CUBIC METRES', NOT 'METRES CUBED'.

Name:

Due date:

Parent's signature:

Part A	/ 8 marks
Part B	/ 8 marks
Part C	/ 8 marks
Part D	/ 8 marks
Total	/ 32 marks

PART A: MENTAL MATHS

🖩 Calculators not allowed

1 Expand $(3x - 1)(6 - x)$.

_____

2 Find the mean of this set of data.

$$-5 \quad 3 \quad 8 \quad -4 \quad 2 \quad 7 \quad 0 \quad 3$$

_____

3 Write 4.09×10^{-3} in decimal form.

_____

4 Two rectangles have lengths in the ratio 2 : 3. If the area of the larger rectangle is 135 cm², find the area of the smaller rectangle.

_____

5 A survey asked 50 students whether they have travelled to Korea or China. The results were: Korea 28, China 22, both Korea and China 5.

a Complete the Venn diagram to show this information.

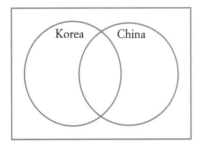

b How many people have not visited either country?

_____

c How many people have visited Korea but not China?

_____

d What is the probability that a person selected randomly from this group has not been to China?

_____

PART B: REVIEW

Find the surface area of each solid, correct to one decimal place for questions **3–4** (2 marks each).

1

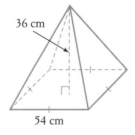

36 cm

54 cm

_____

_____

_____

_____

2

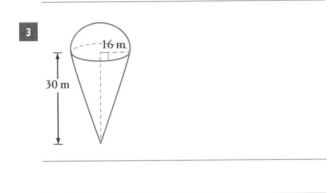

18 cm

12 cm

_____

_____

_____

_____

3

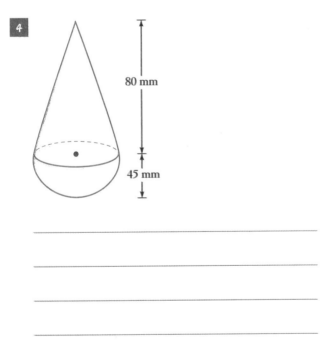

16 m

30 m

_____

_____

_____

4

80 mm

45 mm

_____

_____

_____

_____

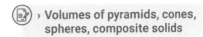 › Volumes of pyramids, cones, spheres, composite solids

Find the volume of each solid, correct to the nearest whole number. (2 marks each)

1

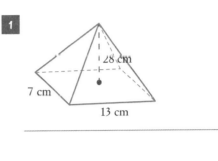

28 cm

7 cm

13 cm

_____

_____

_____

_____

2

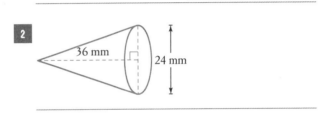

36 mm

24 mm

_____

_____

_____

_____

3

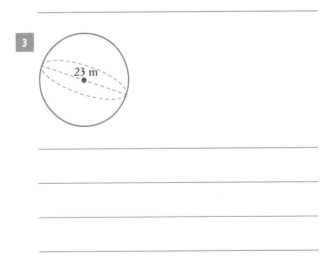

23 m

_____

_____

_____

_____

4

13 cm

12 cm

5 cm

_____

_____

_____

_____

PART D: NUMERACY AND LITERACY

1 Explain what the formula $V = \frac{1}{3}Ah$ does, including what A and h stand for.

_____

_____

_____

_____

2 Write the formula for the volume of a sphere.

_____

3 Find, correct to 2 decimal places, the volume of each solid.

a

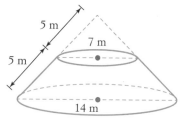

5 m

5 m

7 m

14 m

_____

_____

_____

_____

b

57 mm

|← 35 mm →|

_____

_____

_____

c

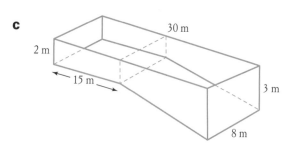

30 m

2 m

15 m

3 m

8 m

_____

_____

_____

_____

STARTUP ASSIGNMENT 5 ⑤

BEFORE WE TACKLE THE HARDER ALGEBRA IN THE 'PRODUCTS AND FACTORS' TOPIC, LET'S REVISE THE BASICS.

PART A: BASIC SKILLS

/ 15 marks

1 Evaluate $2x^2 - 6$ if $x = -1$. _____ _____

2 Divide $5100 between Kath and Kim in the ratio 11 : 6.

_____

_____

_____

3 Expand and simplify:

$m(3m - 4) + 2m(5 - m)$ _____

_____

_____

4 Find the size of one angle in a regular octagon.

_____

5 Write 0.000 007 46 in scientific notation.

_____

6 Find the median of these numbers:

3, 8, 5, 12, 7, 2

_____

7 A rhombus is a square. True or false?

_____

8 What is the probability that the next 2 children born are both girls?

_____

9 Calculate, correct to 2 decimal places, the surface area of this cylinder.

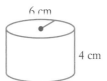
6 cm
4 cm

_____

10 Solve $\dfrac{2x - 5}{4} = 7$. _____

_____

11 Find the value of k in the diagram below.

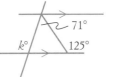
71°
$k°$
125°

_____

12 Graph the line $x = 4$ on a number plane.

13 After an 11% discount, a jacket sells for $40.05. What was its original price?

_____

_____

14 a Find the size of ∠C in the diagram below, correct to the nearest degree.

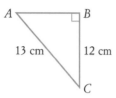

b Calculate the area of the triangle.

PART B: ALGEBRA   / 25 marks

15 List the factors of:

a 9 _____

b 51 _____

16 Simplify:

a $3m \times 7m^2$ _____

b $\dfrac{1}{2}x \times 10$ _____

c $4k \times 3k^3$ _____

d $4g \times (-2g)$ _____

e $8p \times 2q$ _____

f $(3u^4)^2$ _____

17 Find the highest common factor of:

a 12 and 20 _____

b $40xy$ and $24y$ _____

18 List the square numbers from 1 to 100.

_____

_____

19 Expand:

a $3(x + 7)$ _____

b $2(5m - n)$ _____

c $4(1 - 3g)$ _____

d $-2(6p + 5)$ _____

e $m(m + 7)$ _____

f $p(3p - 4)$ _____

g $3k(2k - 4)$ _____

h $-5y(3 - y)$ _____

20 Factorise:

a $3x + 12$ _____

b $10m - 20$ _____

c $3p^2 - 5p$ _____

d $-8m + 18$ _____

e $-14y^2 - 8y$ _____

PART C: CHALLENGE   Bonus   / 3 marks

You are given 12 coins that look identical, but one is a counterfeit (fake) and weighs *less* than the others.

You are given a balance scale. Can you find the counterfeit coin after only 3 weighings?

9780170454568

IF YOU GET STUCK ON THESE, ASK A TEACHER OR FRIEND FOR HELP.

Simplify each expression.

1 $\dfrac{4x}{10} + \dfrac{2x}{10}$

2 $\dfrac{3y}{5} + \dfrac{2y}{5}$

3 $\dfrac{3d}{2} + \dfrac{d}{4}$

4 $\dfrac{5h}{6} + \dfrac{h}{2}$

5 $\dfrac{h}{5} + \dfrac{4h}{3}$

6 $\dfrac{9k}{7} + \dfrac{5k}{2}$

7 $\dfrac{4x}{3} + \dfrac{5x}{8}$

8 $\dfrac{4}{2x} + \dfrac{6}{2x}$

9 $\dfrac{11}{3d} + \dfrac{7}{d}$

10 $\dfrac{7y}{2} - \dfrac{4y}{2}$

11 $\dfrac{10p}{3} - \dfrac{4p}{3}$

12 $\dfrac{10r}{4} - \dfrac{3r}{2}$

13 $\dfrac{6e}{10} - \dfrac{e}{5}$

14 $\dfrac{13d}{6} - \dfrac{3d}{4}$

15 $\dfrac{9p}{7} - \dfrac{p}{3}$

16 $\dfrac{14}{5d} - \dfrac{10}{d}$

17 $\dfrac{4r}{3} + \dfrac{5r}{3} - \dfrac{6r}{3}$

18 $\dfrac{g}{2} + \dfrac{3g}{4} - \dfrac{6g}{4}$

19 $\dfrac{2e}{5} \times \dfrac{5y}{6}$

20 $\dfrac{4r}{3} \times \dfrac{6t}{3}$

21 $\dfrac{9y}{2} \times 10$

22 $8 \times \dfrac{5g}{2}$

23 $\dfrac{6f}{4} \times \dfrac{20}{3}$

24 $\dfrac{25h}{12} \times \dfrac{16g}{10}$

25 $\dfrac{22d}{10e} \times \dfrac{15e}{18}$

26 $\dfrac{14r}{20} \times \dfrac{6r}{7}$

27 $\dfrac{10e}{14d} \times \dfrac{10e}{14d}$

28 $\dfrac{5d}{2} \div \dfrac{25}{3}$

29 $\dfrac{20a}{6} \div \dfrac{4a}{3}$

30 $\dfrac{30f}{4} \div \dfrac{10}{7e}$

31 $\dfrac{12ad}{9c} \div \dfrac{3d}{4}$

32 $\dfrac{16xy}{10} \div \dfrac{6x}{15y}$

33 $\dfrac{24}{6g} \div \dfrac{8g}{h}$

34 $\dfrac{9cd}{4e} \div \dfrac{15c}{14e}$

35 $\dfrac{36mn}{5} \div 4$

36 $\dfrac{28h}{16f} \div 6$

⑤ FACTORISING PUZZLE

> MATCH EACH EXPRESSION WITH ITS FACTORISATION NEXT PAGE TO DECODE THE RIDDLE.

PUZZLE SHEET

| 25 | 24 | 7 | 5 |

| 10 | 27 |

| 18 | 24 | 16 |

| 1 | 10 | 30 | 30 | 16 | 3 | 26 | 21 | 28 | 16 |

| 11 | 16 | 18 | 25 | 16 | 26 | 21 |

| 14 |

| 1 | 22 | 28 | 5 | 22 | 3 |

| 7 | 21 | 1 |

| 14 | 21 |

| 7 | 4 | 17 | 26 | 11 | 3 | 14 |

| 12 | 5 | 13 | 1 | 16 | 21 | 18 | ?

| 18 | 24 | 16 |

| 1 | 22 | 28 | 5 | 22 | 3 |

| 3 | 26 | 28 | 5 | 10 | 30 | 10 | 16 | 27 |

| 13 | 12 |

| 25 | 24 | 10 | 4 | 26 |

| 5 | 24 | 16 |

| 27 | 18 | 13 | 1 | 26 | 21 | 5 |

| 30 | 7 | 28 | 18 | 22 | 3 | 10 | 27 | 16 | 12 | .

The numbers in the grid above match the question numbers. Factorise each expression and match it with an answer from the 'Key' provided on the next page. Fill in the grid above with the letters that match the questions, to decode a riddle.

1 $3x^2 + 3x - 6$

_____

2 $2x^2 - x - 10$

_____

3 $x^2 + 5x + 4$

_____

4 $x^2 - 64$

_____

5 $2x^2 + 7x - 4$

_____

6 $3x^2 - 6x + 21$

_____

7 $x^2 - 3x - 28$

_____

8 $5x^2 - 10x - 120$

_____

9 $-2x^2 + 72$

_____

10 $-3x^2 + 12x - 9$

_____

11 $3d^2 - 24d + 48$

_____

12 $d^2 - 7d + 10$

_____

13 $d^2 - 16d + 64$

_____

14 $4d^2 - 4d - 15$

_____

15 $2d^2 - 6d - 20$

_____

16 $2d^2 - 13d - 7$

_____

17 $4d^2 + 6d - 14$

_____

18 $2d^2 - 9d + 10$

_____

19 $4d^2 - 24d + 36$

_____

20 $d^2 - 100$

_____

21 $n^2 + 3n - 54$

_____

22 $5n^2 + 8n + 3$

_____

23 $3n^2 - 8n - 16$

_____

24 $2n^2 - 36n + 162$

_____

25 $n^2 - 11n + 24$

_____

26 $8n^2 - 2$

_____

27 $3n^2 + 7n - 6$

_____

28 $5n^2 + 35n + 60$

_____

29 $4n^2 - 16$

_____

30 $48n^2 - 75$

_____

Key

A	$(x - 7)(x + 4)$	**I**	$-3(x - 1)(x - 3)$	**S**	$(d - 5)(d - 2)$
A	$(2d - 5)(2d + 3)$	**J**	$4(d - 3)^2$	**S**	$(n + 3)(3n - 2)$
B	$3(d - 4)^2$	**K**	$(2x - 5)(x + 2)$	**T**	$(x + 4)(2x - 1)$
C	$5(n + 3)(n + 4)$	**L**	$(x + 8)(x - 8)$	**T**	$(2d - 5)(d - 2)$
D	$3(x + 2)(x - 1)$	**M**	$2(d + 2)(d - 5)$	**U**	$(d - 8)^2$
E	$(2d + 1)(d - 7)$	**N**	$(n - 6)(n + 9)$	**V**	$4(n + 2)(n - 2)$
E	$2(2n + 1)(2n - 1)$	**O**	$(5n + 3)(n + 1)$	**W**	$(n - 3)(n - 8)$
F	$3(4n + 5)(4n - 5)$	**P**	$(3n + 4)(n - 4)$	**X**	$3(x^2 - 2x + 7)$
G	$2(2d^2 + 3d - 7)$	**Q**	$-2(x + 6)(x - 6)$	**Y**	$(d + 10)(d - 10)$
H	$2(n - 9)^2$	**R**	$(x + 1)(x + 4)$	**Z**	$5(x + 4)(x - 6)$

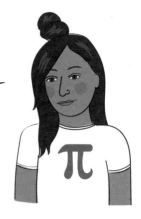

EVERY MATHS TOPIC HAS ITS OWN TERMINOLOGY. HOW WELL DO YOU KNOW THE LANGUAGE OF ALGEBRA?

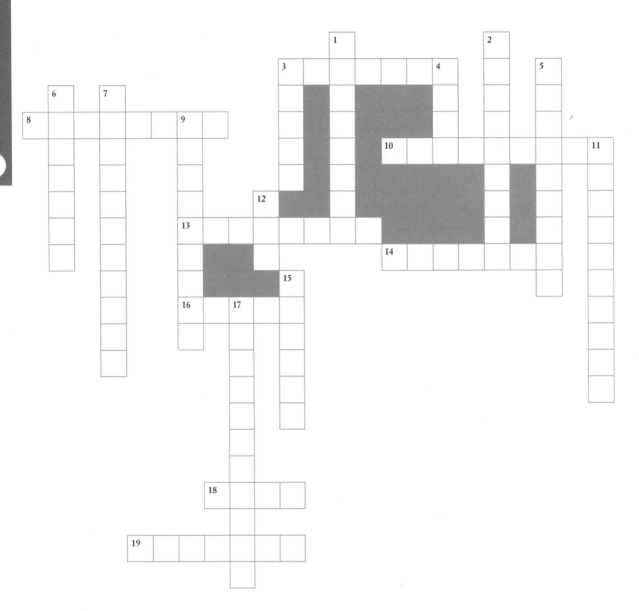

Clues across

3 The answer to a multiplication

8 An algebraic expression with 2 terms, for example $x + 9$, $2y - 12$.

10 The number at the top of a fraction

13 ( )

14 The plural of index

16 $\dfrac{a^m}{a^n} = a^{m-n}$ is an _____ law

18 In 5^3, 5 is called the _____

19 The _____ of 12 are 1, 2, 3, 4, 6 and 12

Clues down

1 In $x^2 - 2x + 9$, 9 is called the _____ term

2 $x^2 - 2x + 9$ is an example of a _____ expression

3 In 5^3, 3 is called the _____

4 When expanding an algebraic expression, multiply each _____ inside the brackets by the _____ outside the brackets

5 To take out the HCF of an expression and write it with brackets

6 The H in HCF

7 The number in front of a variable, such as -4 in $-4x$

9 $\dfrac{x}{2}$ is an example of an _____ fraction

11 Raising a number to a power of -1 gives its _____

12 Abbreviation for 'highest common factor'

15 To remove the brackets of an algebraic expression

17 The number at the bottom of a fraction

⑤ INDEX LAWS 2

INDEX MEANS 'POWER', SO INDEX LAWS ARE JUST RULES ABOUT POWERS.

Name:

Due date:

Parent's signature:

Part A	/ 8 marks
Part B	/ 8 marks
Part C	/ 8 marks
Part D	/ 8 marks
Total	/ 32 marks

PART A: MENTAL MATHS

🚫 Calculators not allowed

1 Evaluate 7^{-2}. _____

_____

2 (2 marks) A rectangle is twice as long as it is wide. If its perimeter is 60 cm, find its dimensions.

_____

_____

_____

3 Describe the shape of this statistical distribution.

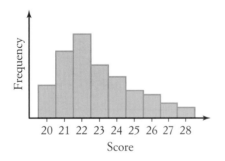

_____

4 Find the area of this shape.

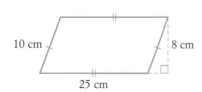

5 Find the x- and y-intercepts of the graph of $y = -4x + 2$.

_____

_____

6 (2 marks) 2 coins are tossed together. What is the probability that at least one of the coins shows heads?

_____

_____

PART B: REVIEW

Simplify each expression.

1 $\left(\dfrac{2}{3}\right)^{-2}$ _____

2 $\left(1\dfrac{2}{5}\right)^{-3}$ _____

3 $\left(\dfrac{5d^3}{3p^4}\right)^{-2}$ _____

4 $3m^2p^{-2}$ _____

5 $(4h)^{-2}$ _____

6 $\left(\dfrac{1}{2}\right)^0 + \dfrac{1}{2}y^0$ _____

7 $\left(\dfrac{1}{2}\right)^0 + \left(\dfrac{1}{2}y\right)^0$ _____

8 $(3x^3)^3 \div x^9$ _____

_____

PART C: PRACTICE

> Fractional indices

1 Write each expression using a fractional index.

   a $\sqrt[10]{100x}$ _____

   b $\sqrt[3]{2mn}$ _____

   c $\sqrt[6]{y^{10}}$ _____

   d $\dfrac{1}{\sqrt[4]{p^5}}$ _____

2 Simplify each expression.

   a $\left(16a^2b^6\right)^{\frac{1}{2}}$ _____

   b $35x \div 5x^{\frac{1}{3}}$ _____

   c $\left(64x^4y^{10}\right)^{\frac{3}{2}}$ _____

   d $\left(1000m^3n^6\right)^{-\frac{2}{3}}$ _____

PART D: NUMERACY AND LITERACY

1 Explain what happens to a number when it is raised to a power of:

   a $\dfrac{1}{3}$

   _____

   _____

   b $\dfrac{1}{6}$

   _____

   _____

2 Write each expression in radical (root) form.

   a $\left(a^m\right)^{\frac{1}{n}}$ _____

   b $\left(a^{\frac{1}{n}}\right)^m$ _____

3 (2 marks) Evaluate $36^{-\frac{3}{2}}$, showing working.

   _____

   _____

   _____

4 Simplify each expression.

   a $\left(125d^{15}\right)^{\frac{4}{3}}$ _____

   b $\sqrt[3]{-64n^{12}}$ _____

5 ALGEBRAIC FRACTIONS

THE RULES FOR WORKING WITH ALGEBRAIC FRACTIONS ARE THE SAME AS THE RULES FOR NORMAL FRACTIONS.

Part A	/ 8 marks
Part B	/ 8 marks
Part C	/ 8 marks
Part D	/ 8 marks
Total	/ 32 marks

HOMEWORK

PART A: MENTAL MATHS

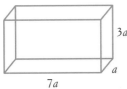 Calculators not allowed

1 Find a simplified algebraic expression for the volume of this rectangular prism.

(rectangular prism with dimensions $7a$, a, $3a$)

_____

2 (2 marks) Expand $(10x - 2y)(10x + 2y)$.

_____

_____

3 Write the equation of the line that has a gradient of -2 and a y-intercept of 3.

_____

4 A card is chosen at random from a set of cards numbered 0 to 9. Find the probability of selecting:

a 7 _____

b a number less than 5 _____

c a number 6 or more _____

5 Write tan θ for this triangle.

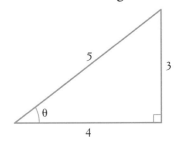

_____

PART B: REVIEW

1 Evaluate without a calculator:

a $\dfrac{2}{3} - \dfrac{3}{7}$ _____

b $\dfrac{4}{9} + \dfrac{5}{6}$ _____

c $\dfrac{6}{7} \times \dfrac{5}{8}$ _____

d $\dfrac{5}{9} \div \dfrac{5}{6}$ _____

2 Simplify each expression.

a $\dfrac{3m^2 n^5}{18mn^2}$ _____

b $2t^{-1}$ _____

c $(5x^3 y^2)^2$ _____

d $(-3ab^2)^4$ _____

PART C: PRACTICE

📝 › Algebraic fractions

Simplify each expression.

1 $\dfrac{4x}{15} + \dfrac{3x}{10}$ _____

2 $\dfrac{y}{16} - \dfrac{y}{24}$ _____

3 $\dfrac{m}{5} - \dfrac{m}{7}$ _____

4 $\dfrac{y}{4} + \dfrac{2y}{5}$ _____

5 $\dfrac{20y}{3x} \times \dfrac{9y}{5x}$ _____

6 $\dfrac{4mn}{9} \times \dfrac{3m}{16n}$ _____

7 $\dfrac{8a}{7} \div \dfrac{40a^2}{3p}$ _____

_____

8 $\dfrac{6mn}{5x} \times \dfrac{4x}{n} \div 18mn$ _____

_____

PART D: NUMERACY AND LITERACY

1 (5 marks) Complete:

a To add or subtract fractions, convert them (if needed) so that they will have the same _____, then add or subtract the _____.

b To multiply fractions, cancel any common factors, then multiply the _____ and _____ separately.

c To divide by a fraction $\dfrac{a}{b}$, multiply by its _____ $\dfrac{b}{a}$.

2 Simplify $\dfrac{3x - 15}{7} \times \dfrac{14x}{x^2 - 5x}$.

_____

_____

_____

3 Explain in words how to simplify:

a 3^{-2} _____

_____

_____

b 3^0 _____

_____

_____

5 SPECIAL BINOMIAL PRODUCTS

ALGEBRA IS ALL ABOUT NOTICING PATTERNS AND RULES. IF YOU CAN TRAIN YOURSELF TO SEE THE PATTERNS, THEN YOU'LL BE GOOD AT MATHS.

Name:

Due date:

Parent's signature:

Part A	/ 8 marks
Part B	/ 8 marks
Part C	/ 8 marks
Part D	/ 8 marks
Total	/ 32 marks

PART A: MENTAL MATHS

🖩 Calculators not allowed

1 Evaluate $(-3)^3$.

_____

2 Test whether {20, 21, 29} is a Pythagorean triad.

_____

_____

_____

_____

3 Evaluate $8 - 3b^2$ if $b = -2$.

_____

4 Convert $\dfrac{1}{6}$ to a recurring decimal.

_____

_____

5 Write the bearing of B from A.

6 Write a simplified algebraic expression for the perimeter of the shape.

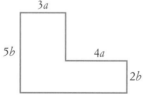

_____

7 Find m and y.

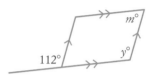

_____

_____

PART B: REVIEW

1 (4 marks) Expand each expression.

a $(p + 4)(p - 1)$

_____

_____

b $(y - 9)(y - 3)$

_____

_____

2 (4 marks) Factorise each expression.

a $x^2 + 8x - 33$

_____

_____

b $a^2 - 11a + 28$

_____

_____

9780170454568

PART C: PRACTICE

1 (4 marks) Expand each expression.

a $(x - 8)^2$

b $(2k + 5)^2$

2 (4 marks) Factorise each expression.

a $2ab + 2da - 3bc - 3dc$

b $6d^2 + d - 2$

c $4y^2 - 81$

PART D: NUMERACY AND LITERACY

1 Write an algebraic expression that is:

a a perfect square

b a difference of 2 squares

2 (4 marks) Expand each expression.

a $(3y - 7)(3y + 7)$

b $(3y - 7)^2$

3 (2 marks) Factorise $3c^2 - 12c - 36$.

HW HOMEWORK

⑥ STARTUP ASSIGNMENT 6

> LET'S GET READY FOR THE COMPARING DATA TOPIC BY REVISING OUR STATISTICS SKILLS.

PART A: BASIC SKILLS   / 15 marks

1 Simplify $(5a)^3$. _____

2 Which size pack of ice cream below gives better value for money?

    A 750 g for $5.20

    B 1.25 kg for $8.60

_____

3 Find the median of these numbers:

8, 8, 7, 4, 5, 4, 3, 6, 6, 7 _____

4 Simplify $\dfrac{6x - 4xy}{2x}$. _____

5 Name this solid.

 _____

6

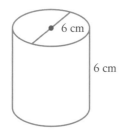

For this cylinder, find to 2 decimal places:

    a its volume _____

    b its surface area. _____

7 Write 62.5% as a fraction. _____

8 Find k if 8 more than triple k is 2 less than 4 times k. _____

_____

9 Complete: 3.6 hours = 3 h _____ min

10 Solve $\dfrac{d}{5} = \dfrac{16}{3}$. _____

_____

11 What is the y-intercept of the graph of $y = -2x$? _____

12 The value of a house increased from $440 890 to $491 600. Calculate the percentage increase in value, correct to one decimal place.

_____

13 A rectangle has a perimeter of 450 cm. Find its length and width if they are in the ratio 5 : 4.

_____

_____

14 A car has a fuel consumption of 8.1 L/100 km. How many whole kilometres can it travel on 45.4 L of fuel? _____

_____

PART B: STATISTICS   / 25 marks

15 This dot plot displays the results of a survey about the number of TVs in homes.

    a How many homes were surveyed?

_____

    b Mode = _____

    c Range = _____

d Mean (to 2 decimal places) = _____

e Median = _____

f Draw a frequency polygon for the data.

16 What is the difference between a **sample** and a **census**?

_____

_____

_____

_____

17 This stem-and-leaf plot shows masses (in kilograms) of a group of students.

Stem	Leaf
5	2 4 5 7 9
6	0 3 3 4 4 4 6 7 8 8
7	0 1 1 2 3 6

Complete the following:

a Mode = _____

b Range = _____

c Median = _____

d How many students weighed less than 65 kg?

_____

e Mean = _____ (to 2 decimal places)

18 This data is from a survey about the number of people living in individual houses.

4 2 1 4 3 5 8 6 3 5

3 1 2 5 4 3 2 7 6 4

5 3 2 4 4 2 4 4 5 3

a Complete the cumulative frequency table.

Score	f	cf
1		
2		
3		
4		
5		
6		
7		
8		

b Mode = _____

c Median = _____

d Range = _____

e How many houses had 3 or fewer people living in them? _____

PART C: CHALLENGE   Bonus   / 3 marks

64 players entered a tennis knockout competition. After each round, only the winners continue playing; the losers are 'knocked out'. How many matches are required to find the winner of the competition? How many rounds are required?

_____

6 LINE OF BEST FIT

YOU'LL NEED A RULER TO DRAW
YOUR LINE OF BEST FIT.

A researcher recorded data on a region's annual rainfall and number of bushfires in the following table.

	A	B	C	D	E	F	G	H
Rainfall (cm)	19	23	24	7	22	27	20	16
Number of bushfires	24	9	12	32	18	10	21	20

1 Plot the data points on the grid below. Label each point with the appropriate letter (A to H).

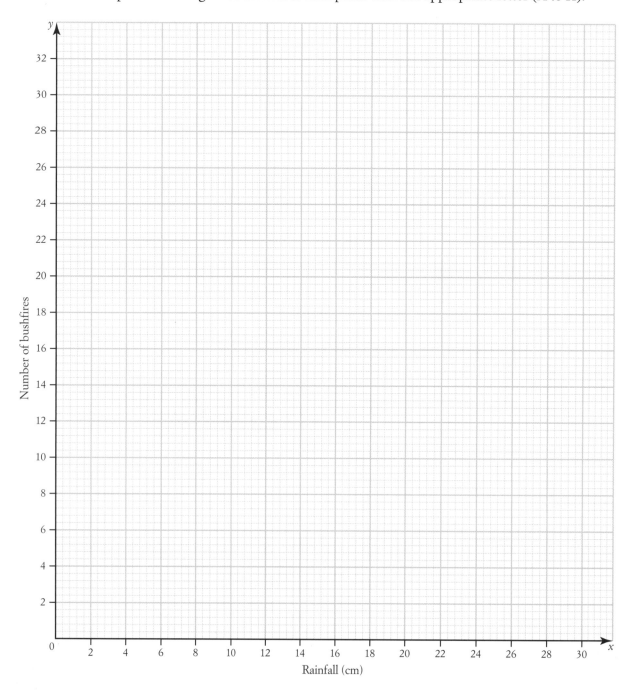

2 Construct a line of best fit.

3 Find the equation of the line of best fit.

_____

_____

_____

_____

_____

4 Use the equation to find the number of bushfires if rainfall is 14 cm.

_____

5 Use the graph to:

   a interpolate the number of bushfires when the rainfall is 10 cm

   _____

   b extrapolate the number of bushfires when the rainfall is 30 cm

   _____

   c estimate the rainfall if there have been 15 bushfires.

   _____

6 What are the limitations of using this line of best fit to predict the number of bushfires, given the rainfall?

_____

_____

_____

_____

6 DATA CROSSWORD

HERE'S A CLUE: ANOTHER NAME FOR AVERAGE IS 'MEASURE OF LOCATION'.

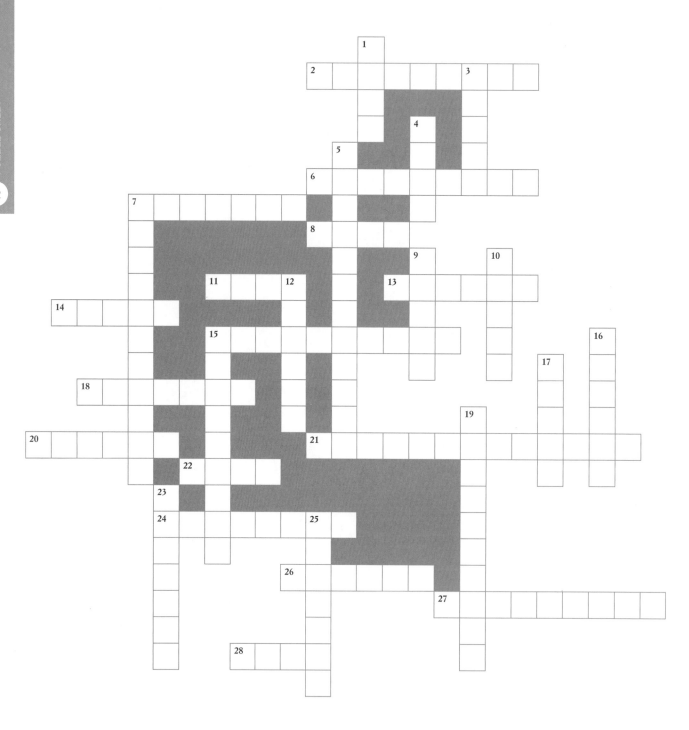

Clues across

2 Q_1, Q_2, Q_3 are called these

6 Frequency column graph

7 A graph of points on a number plane is called a _____ plot.

8 _____-and-leaf plot

11 Something misleading that causes a sample or graph to not truly represent a situation

13 Interquartile range is a measure of _____

14 To find the median, first sort the scores in _____

15 A boxplot illustrates a _____-_____ summary

18 Box-and-_____ plot

20 The average of the 2 middle values if there is an even number of values in a data set

21 This range is the difference between Q_3 and Q_1

22 Sum of the data values divided by the number of values

24 Mean, median, mode are measures of _____ (or central tendency).

26 A scatterplot of points that approximate a line indicates a _____ relationship between 2 variables

27 Data in the form of ordered pairs that measure 2 variables

28 Most frequent value in a data set

Clues down

1 Another word for information

3 Q_1 is called the _____ quartile

4 A simple graph is the dot _____

5 'D' word meaning set of values

7 Not skewed

9 The highest value is also called the _____ extreme

10 The simplest measure of spread

12 A 'twisted' distribution of data values

15 How often a value appears in a data set

16 The median represents the _____ value

17 The interquartile range is the range of the middle _____ %

19 'cf' stands for _____ frequency

23 Describes where many of the values in a data set are grouped

25 An extreme data value that is much different to the others

⑥ BOXPLOTS

IN A BOXPLOT, THE LENGTH OF THE BOX IS THE INTERQUARTILE RANGE. AGREE?

HW HOMEWORK

PART A: MENTAL MATHS

🚫 Calculators not allowed

1 Evaluate each expression if $a = -3$, $b = 2$ and $c = 5$.

a $\sqrt{a+b+c}$ _____

b $\dfrac{a}{c-b}$ _____

2 (2 marks) Simplify $\dfrac{6wy}{zx} \div \dfrac{10y}{xz} \div \dfrac{5z}{w}$.

_____

3 For $\triangle ABC$, find the value of:

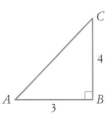

a AC _____

b $\cos C$ _____

4 (2 marks) Find x, giving reasons.

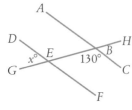

_____

_____

PART B: REVIEW

For this data set, find:

Stem	Leaf
1	0 3 6
2	1 6 7 8
3	5 5 6
4	1 1 5 6 9
5	0 3 6 8

1 the range _____

2 the median _____

3 the mean (to 2 decimal places)

_____

4 the lowest value _____

5 the lower quartile (Q_1) _____

6 the upper quartile (Q_3) _____

7 the highest value _____

8 the interquartile range _____

PART C: PRACTICE

 › Boxplots

1 For this data set, find:

```
        •
        •     •
        •     •           •
        •     •     •      •
        •     •     •      •     •
  •     •     •     •      •     •           •
  •     •     •     •      •     •     •     •     •
  ├─────┼─────┼─────┼─────┼─────┼─────┼─────┼─────┼─────┼─────┤
  0     1     2     3     4     5     6     7     8     9     10
```

a the lowest value _____

b the lower quartile _____

c the median _____

d the upper quartile _____

e the highest value _____

f the interquartile range _____

2 (2 marks) Represent the data from question **1** on a boxplot.

PART D: NUMERACY AND LITERACY

1 (2 marks) Complete each box with a percentage.

2 On a boxplot, what value is represented:

a on the end of the right whisker?

_____

b on the left edge of the box?

_____

c on the vertical bar inside the box?

_____

3 (3 marks) Name the 5 things found in a five-number summary.

_____

_____

_____

_____

_____

HW HOMEWORK

6 COMPARING DATA

ARE YOU GETTING THE HANG OF STATISTICS YET? NEWS REPORTS SHOW DATA AND GRAPHS EVERY DAY.

Name:

Due date:

Parent's signature:

Part A	/ 8 marks
Part B	/ 8 marks
Part C	/ 8 marks
Part D	/ 8 marks
Total	/ 32 marks

PART A: MENTAL MATHS

 Calculators not allowed

1 For this triangle, find the value of:

$$\text{17 (hypotenuse), } x, \theta, 15$$

a x _____

b $\cos\theta$ _____

c $3\tan\theta$ _____

2 (2 marks) The price of a computer monitor was reduced by 20%. Find the original price of the monitor if its sale price was $160.

_____

_____

3 Simplify:

a $(-3ab^2)^3$ _____

b $3a^{-1}b^0$ _____

_____

4 Solve $3x + 4 = 5x - 9$.

_____

_____

_____

PART B: REVIEW

1 For this boxplot, find:

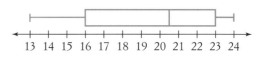

13 14 15 16 17 18 19 20 21 22 23 24

a the median _____

b the lowest value _____

c the upper quartile _____

d the interquartile range _____

2 (4 marks) Draw a boxplot for the data below, showing all values of the five-number summary.

9	2	12	9	12	5	4
2	8	8	1	6	2	10

PART C: PRACTICE

> Parallel boxplots
> Comparing data sets

1 (5 marks) These parallel boxplots show the heights (cm) of boys and girls in a Year 2 class.

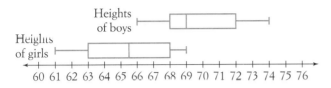

a Complete this table:

	Median	Interquartile range
Boys		
Girls		

b State one difference between the heights of boys and girls in this class.

_____

_____

2 The ages of males and females at a local library are shown in the stem-and-leaf plot.

Male		Female
	0	5 7 8
4 3 0	1	0 2 5 6 6 8
7 7 6	2	7
4 2	3	6
	10	3

a Find the mean for the males.

_____

b Find the mean for the females.

_____

c State one difference between the ages of males and females at the library.

_____

PART D: NUMERACY AND LITERACY

These back-to-back histograms show the ages of teachers at Ashfield and Burwood high schools.

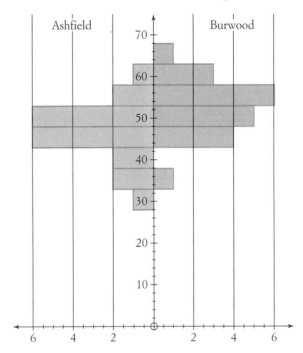

1 (5 marks) Complete this table:

Age group	Ashfield Frequency	Burwood Frequency
28– < 33	1	0
33– < 38		
38– < 43		
43– < 48		
48– < 53		
53– < 58		
58– < 63		
63– < 68		

2 Describe the shape of each set of data.

Ashfield: _____

Burwood: _____

3 State one difference between the ages of teachers at the 2 schools.

_____

_____

⑥ LINE OF BEST FIT

ALGEBRA MEETS STATISTICS HERE AS A LINE OF BEST FIT IS USED TO REPRESENT THE DATA ON A SCATTERPLOT.

PART A: MENTAL MATHS

🚫 Calculators not allowed

1 (2 marks) Find the x- and y-intercepts of the graph of $3y - 4x - 2 = 0$.

2 Write the formula for the volume of a cone.

3 For the interval joining $(-2, 4)$ and $(6, 3)$, find:

a its length in exact form

b its midpoint

c its gradient

4 Find $12\frac{1}{2}\%$ of \$620.

5 Factorise $\dfrac{x^2}{16} - \dfrac{y^2}{49}$.

PART B: REVIEW

1 (5 marks) The heights and armspans of a group of swimmers are shown in the table.

Height (h cm)	Arm span (s cm)
190	192
192	194
199	202
188	189
197	200
187	188
194	196
198	202
193	197
199	200

9780170454568

a Plot the data in a scatterplot.

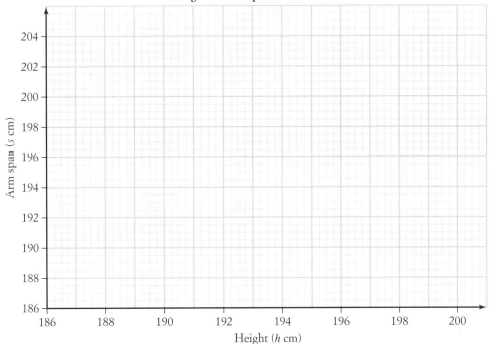

Height vs Arm span of male swimmers

b Describe the strength and direction of the relationship between the swimmers' height and arm span.

_____

c According to the graph, is arm span considered to be the dependent variable or the independent variable?

_____

2 Describe the strength and direction of the relationship shown in each scatterplot.

a

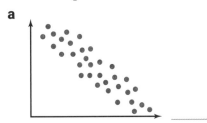

_____

b

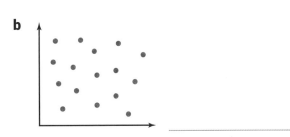

_____

c

_____

PART C: PRACTICE

> › Standard deviation
> › Line of best fit

1 (4 marks)

a Draw a line of best fit on the scatterplot from Part B, question **1**.

b Find the equation of the line of best fit.

_____

_____

_____

_____

_____

2 Find, correct to 2 decimal places, the standard deviation of each set of data.

a 38 17 42 36 38 41 19 22 54

_____

b

Stem	Leaf
5	0 1 2
6	3 4
7	1 5
8	6
9	1 3

_____

c

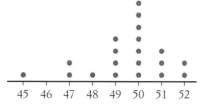

_____

PART D: NUMERACY AND LITERACY

1 Write one feature of a line of best fit for a scatterplot.

_____

_____

2 What is predicting the values of a graph beyond the points on a scatterplot called?

_____

3 Complete: The standard deviation is a measure of _____ whose value is calculated using _____ value in a data set.

4 (4 marks) The outdoor temperature and ice cream sales of a store are shown on this scatterplot.

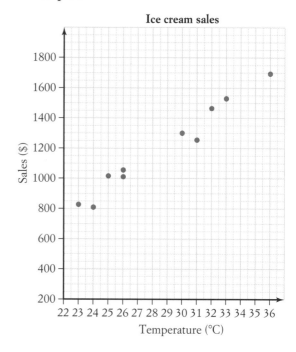

a Construct a line of best fit on the graph.

b Find the equation of the line of best fit.

_____

_____

_____

_____

c Use the graph to estimate the sales of ice cream for a temperature of 27°C.

_____

_____

STATISTICAL CALCULATIONS ⑥

FOR STANDARD DEVIATION, YOU WILL NEED THE STATISTICS MODE OF YOUR CALCULATOR.

Complete the table for each data set (**A** to **J**) below, giving values to two decimal places where appropriate.

Data set	Mean	Mode	Median	Range	Interquartile range	Standard deviation
A						
B						
C						
D						
E						
F						
G						
H						
I						
J						

A 13 16 15 14 16 15 13 12
    11 15 14 14 14 18 11

B 34 40 28 55 27 30
    14 25 71 36 42 52

C

Score, x	Frequency, f	Cumulative frequency
7	2	
8	5	
9	8	
10	6	
11	3	
12	1	

D

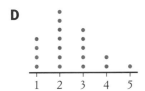

E

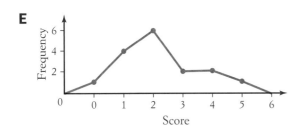

F

Stem	Leaf
5	3 3 4 9
6	0 2 5 7
7	1 1 4 5 6 6 8 9
8	3 4 4 6 8 9
9	2 6 8

G 0 6 5 3 8 7 4 5

H Add 5 to each score in set G to make a new set of scores. Predict how each of the statistics will be affected by this change, and then investigate your predictions.

I Multiply each score in set G by 3 to make a new data set. Predict how each statistic will be affected, and then investigate.

J A new score, 20, is combined with the set of scores in G to make a total of nine scores. Predict how the statistics will be affected by the inclusion of this outlier, and then investigate your predictions.

STARTUP ASSIGNMENT 7

THIS ASSIGNMENT WILL HELP YOU
PREPARE FOR THE EQUATIONS
AND LOGARITHMS TOPIC.

PART A: BASIC SKILLS / 15 marks

1 Simplify 3^{-2}. _____

2 Convert 75 km/h to m/s. _____

_____

3 Which is the better buy? _____

  A 250 mL for $1.45

  B 600 mL for $3.60

4 Write the geometrical symbol for

  'is congruent to'. _____

5 Convert $\dfrac{5}{6}$ to a percentage. _____

6 Find the value of p.

102° $p°$

_____

7 A map's scale is 1 : 50 000. What distance in

kilometres does 4 cm on this map represent?

_____

8 $A(-1, 2)$ and $B(4, 6)$ are points on the number

plane. Find:

  a the length of AB in surd form _____

  _____

  b the gradient of AB. _____

9 Write 1 049 720 correct to 3 significant figures.

_____

10 What is the most frequent value in a set of data

called? _____

11 Simplify 27 : 18 : 6.

_____

12

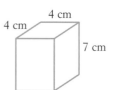

4 cm

4 cm

7 cm

For this square prism, find its:

  a surface area _____

  _____

  b volume. _____

13 Write 0.006 in scientific notation. _____

PART B: EQUATIONS / 25 marks

14 Solve $3p + 4 = 18$. _____

15 $-1 > 6$, true or false? _____

16 If $y = 3x - 4$, find y when $x = -1$. _____

17 Expand:

  a $2(5a - 3)$ _____

  b $-6(2a - 4)$ _____

18 Test whether $x = 4$ is a solution to

$5x + 6 = 3x + 14$. _____

_____

19 Solve $\dfrac{k}{4} = 4$. _____

20 Solve $11 - 2r = 17$. _____

_____

21 Complete this table for $2x + y = 16$.

x	-2	-1	0	1	2	3
y						

22 Solve $\dfrac{2y}{3} + 6 = 10.$ _____

_____

23 Solve $-3m = -25.$ _____

24 Expand and simplify:

a $2(r + 5) + 3(4r - 2)$ _____

_____

b $(t + 9)(t - 1)$ _____

_____

25 Solve $\dfrac{p + 7}{5} = 8.$ _____

_____

26 Solve $4(h - 2) = 30.$ _____

_____

27 Use an equation to find 3 consecutive odd numbers that add to 117.

_____

_____

_____

_____

28 Graph the line $y = 3x - 2$ on a number plane.

29 Find the size of the obtuse angle in this triangle.

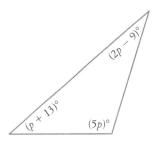

_____

30 If $2x + 3y = 24$ and $y = 6$, find x. _____

_____

_____

31 The cost of hiring a karaoke machine is given by the formula $C = 90 + 25h$, where C is the total cost and h is the number of hours.

a Find the cost of hiring the machine for 6 hours. _____

b How many hours will $165 pay for?

_____

_____

32 Test whether $n = -28$ is the solution to $\dfrac{n - 2}{5} - \dfrac{n}{4} = 1.$ _____

_____

_____

33 Solve $3x - 5 = x + 9.$ _____

_____

_____

PART C: CHALLENGE   Bonus   / 3 marks

A farmer has emus and pigs. Altogether there are 250 creatures and they have 834 legs. How many of each animal is there?

_____

⑦ GRAPHING INEQUALITIES

INEQUALITIES ARE SHOWN ON A NUMBER LINE BY ARROWS AND CIRCLES. SHADE IN THE CIRCLE IF THAT NUMBER IS INCLUDED.

Graph each inequality on the number line provided.

1 $x \geq 3$

2 $x > 1$

3 $x < -2$

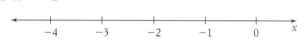

4 $x \leq 5$

5 $x > -1$

6 $x \geq 0$

7 $x \leq -1$

8 $x > 3\frac{1}{2}$

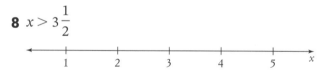

9 $x + 8 \leq 12$

10 $4x + 5 \leq -1$

11 $7x - 2 < -2$

12 $8x \geq -24$

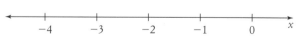

9780170454568

13 $x + 5 < 7$

14 $3x > 15$

15 $8x - 1 \leq 11$

16 $2(x + 5) \geq -2$

17 $4(x + 4) \geq 6$

18 $7(x - 4) < -7$

19 $\dfrac{x}{4} \leq 2$

20 $\dfrac{x + 10}{2} > 3$

21 $2(2x + 9) < 20$

22 $6(x - 7) \geq -9$

23 $\dfrac{x - 8}{6} \leq -2$

24 $\dfrac{2x + 7}{3} > 4$

⑦ LOGARITHMS REVIEW

LOGARITHMS ARE IMPORTANT FOR ADVANCED MATHS IN YEAR 11.

1 Evaluate each expression.

a $\log_{10}1000$ _____

b $\log_{2}256$ _____

c $\log_{5}125$ _____

d $\log_{7}49$ _____

e $\log_{6}\left(\dfrac{1}{6}\right)$ _____

f $\log_{8}\left(\dfrac{1}{64}\right)$ _____

g $\log_{9}3$ _____

h $\log_{3}1$ _____

2 Write each equation in logarithmic form.

a $4^3 = 64$ _____

b $5^2 = 25$ _____

c $2^r = 32$ _____

d $y^6 = z$ _____

3 Write each equation in index form.

a $\log_{10}100 = 2$ _____

b $\log_{2}8 = 3$ _____

c $\log_{3}b = 7$ _____

d $\log_{a}16 = 2$ _____

4 Simplify and evaluate each expression.

a $\log_{3}3^5$ _____

b $\log_{4}32 + \log_{4}2$ _____

c $\log_{10}900 - \log_{10}9$ _____

d $3\log_{6}2 + \log_{6}27$ _____

e $\log_{5}15 + \log_{5}2 - \log_{5}6$ _____

f $3\log_{10}5 - \log_{10}2 + \log_{10}16$ _____

g $\log_{9}12 - 2\log_{9}2$ _____

h $2\log_{2}4 + \log_{2}20 - \dfrac{1}{2}\log_{2}25$ _____

5 Simplify each expression.

a $\log_{a}5 + \log_{a}6$ _____

b $\log_{a}25 - \log_{a}5$ _____

c $\log_{a}6 + \log_{a}8 - \log_{a}4$ _____

d $\log_{a}4a^2$ _____

e $\log_{a}20 - (\log_{a}5 + \log_{a}4)$ _____

f $\log_{a}6a^2 - \log_{a}2a^4$ _____

g $\dfrac{1}{2}\log_{a}16a$ _____

h $\log_{a}3c + \dfrac{1}{3}\log_{a}d - 2\log_{a}e$ _____

9780170454568

6 Expand each expression.

a $\log_a(xy)$ _____

b $\log_a(x^n)$ _____

c $\log_a\left(\dfrac{pq}{r}\right)$ _____

d $\log_a(4s^2)$ _____

e $\log_a\sqrt{xy}$ _____

f $\log_a\left(\dfrac{c}{d^2}\right)$ _____

7 If $\log_n 4 = x$, $\log_n 6 = y$ and $\log_n 10 = z$, express each expression in terms of x, y and z.

a $\log_n 40$ _____

b $\log_n 2$ _____

c $\log_n 160$ _____

d $\log_n 15$ _____

8 If $\log_{10} 6 = 0.7782$, find the value of each expression.

a $\log_{10} 36$ _____

b $\log_{10}\dfrac{1}{6}$ _____

c $\log_{10} 60$ _____

d $\log_{10} 0.6$ _____

e $\log_{10} 360$ _____

f $\log_{10}\sqrt{6000}$ _____

9 Solve each equation, correct to 2 decimal places where necessary.

a $4^x = 4096$ _____

b $6^x = 1\,679\,616$ _____

c $5^x = 120$ _____

d $2^x + 2 = 3264$ _____

e $4^x - 3 = 1984$ _____

f $3^{4-x} = 300$ _____

10 Solve each equation.

a $\log_3 x = 4$ _____

b $\log_5 x = 5$ _____

c $\log_4 x = \dfrac{1}{2}$ _____

d $\log_x 36 = 2$ _____

e $\log_x \dfrac{1}{10} = -1$ _____

f $\log_x 64 = 3$ _____

9780170454568

Chapter 7 Equations and logarithms   **79**

WS WORKSHEET

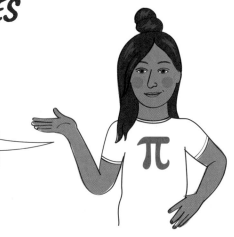

MATHS HAS ITS OWN LANGUAGE AND SOME OF THE CLUES IN THIS PUZZLE ARE QUITE CHALLENGING. WORK WITH A FRIEND ON THIS.

Clues down

1 We must do this to $u^2 + 3u - 28$ before we solve $u^2 + 3u - 28 = 0$

2 < means _____ than

5 [ ]

9 Inequalities can be graphed on a number _____ .

10 The answer to an equation or inequality is called its _____ .

11 Number less than 0

14 Most quadratic equations have _____ solutions

15 An equation involving x^2

18 Abbreviation for 'lowest common multiple'

21 A rule written as an equation

Clues across

3 The variable on the left side of a formula is called its s _____

4 To replace a variable with a number

6 To remove the brackets in an algebraic expression

7 ≥ means _____ than or equal to

8 $3x + 6 \leq 15$ is an example of one

12 To find the answer to an equation or inequality

13 $18 - 5x = 10$ is an example of one

16 $\sqrt{}$ means square _____

17 A letter that stands for a number

19 Equations can be solved by these 'opposite' operations

20 The M in LCM

21 It has a denominator

22 Abbreviation for 'right-hand side'

23 A simple way of solving equations is guess and _____

24 Abbreviation for 'left-hand side'

(7) EQUATIONS AND FORMULAS

NOTICE THE RED QUESTIONS NEXT PAGE? THEY ARE QUITE CHALLENGING AND WILL REQUIRE EXTRA TIME.

Name:

Due date:

Parent's signature:

Part A	/ 8 marks
Part B	/ 8 marks
Part C	/ 8 marks
Part D	/ 8 marks
Total	/ 32 marks

PART A: *MENTAL MATHS*

Calculators not allowed

1 Kathie bought a laptop for $800 and sold it at a profit of 8%.

a What was the profit?

_____

b For what price did Kathie sell the laptop?

_____

2 Factorise:

a $-20mn + 25n^2$

_____

b $-x(3 + y) - 4(3 + y)$

_____

3 (4 marks) Find the value of each variable, giving reasons.

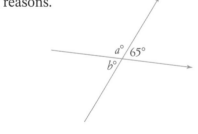

_____

_____

_____

PART B: *REVIEW*

1 (4 marks) Solve each equation.

a $2a = 9(a - 5)$

_____

_____

b $5(3x + 2) = -40$

_____

_____

2 (4 marks)

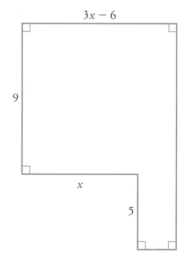

a Find x if the perimeter of this shape is 46 cm.

_____

_____

b Find the area of the shape.

_____

_____

9780170454568

PART C: PRACTICE

📝 › Equation problems
› Equations and formulas

1 The formula $F = \dfrac{9C}{5} + 32$ converts a temperature in °C (Celsius) to °F (Fahrenheit). Convert 25°C to °F.

_____

2 (3 marks) **a** Find x.

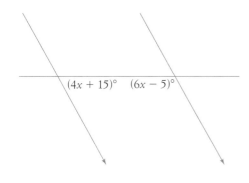

$(4x + 15)°$ $(6x - 5)°$

_____

_____

_____

b Find the size of the acute angle.

_____

3 (4 marks) The surface area of a closed cylinder has the formula $SA = 2\pi r^2 + 2\pi rh$. Calculate, correct to one decimal place:

a the surface area of a cylinder with radius 4.5 cm and height 6.2 cm

_____

b the height of a cylinder with surface area 2104.23 cm² and radius 12.3 cm

_____

_____

_____

PART D: NUMERACY AND LITERACY

1 (2 marks) If 8 more than a number is the same as 3 more than double the number, what is the number?

_____

_____

_____

_____

2 (3 marks) Gianni's bag contained only 10c and 50c coins. He had 124 coins, with a total of $55.60. How many 10-cent coins did he have?

_____

_____

_____

_____

3 (3 marks) The sum of Sanjeev's and her mother's ages is 65. In 3 years, 3 times Sanjeev's age less 21 will be the same as her mother's age. How old is Sanjeev?

_____

_____

_____

_____

7 INEQUALITIES

SOLVING INEQUALITIES IS A NEW SKILL, BUT THE RULES ARE SIMILAR TO THOSE FOR SOLVING EQUATIONS.

PART A: MENTAL MATHS

Calculators not allowed

1 Complete: 1500 cm = _____ m

2 Evaluate $7^2 + 9^2$ _____

3 Write the recurring decimal 0.124124124... using dot notation.

_____

4 For this set of data, find:

22   24   26   24   23   30   31   22

26   27   31   24   25   22   21   29

a the mode

_____

b the median

_____

5 Find the area of each shape.

a

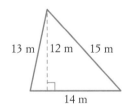

13 m   12 m   15 m

14 m

_____

b

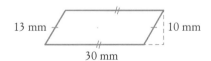

13 mm   10 mm

30 mm

_____

6 Write one property of the diagonals of a parallelogram.

_____

_____

PART B: REVIEW

1 (4 marks) Write each inequality in words and write a number that follows that inequality.

a $x > 4$

_____

_____

b $x \leq 4$

_____

_____

2 Graph each inequality on a number line.

a $x \leq -1$

_____

b $x > 7$

_____

c $x \geq 0$

_____

d $x < 4$

_____

PART C: PRACTICE

> › Graphing inequalities on a number line
> › Soliving inequalities

1 Write the inequality illustrated by each number line.

a

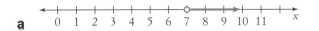

_____

b

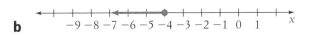

_____

2 (6 marks) Solve each inequality.

a $3m - 11 \leq 10$

_____

_____

b $5(x - 5) > -30$

_____

_____

c $\dfrac{y + 6}{7} \geq -4$

_____

_____

PART D: NUMERACY AND LITERACY

1 What word means the answer to an equation or inequality? _____

2 True or false? $-3 \geq -3$.

_____

3 (2 marks) Complete: When solving inequalities, if you _____ both sides by a negative number, you must _____ the inequality sign.

4 (4 marks) Solve each inequality and graph its solution on a number line.

a $\dfrac{m}{6} \leq -3$

b $-5 - 20x > 5$

⑦ LOGARITHMS 1

LOGARITHMS ARE HARD. IT HELPS IF YOU REMEMBER THAT THE LOG OF A NUMBER IS A POWER.

Name:

Due date:

Parent's signature:

Part A	/ 8 marks
Part B	/ 8 marks
Part C	/ 8 marks
Part D	/ 8 marks
Total	/ 32 marks

PART A: MENTAL MATHS

🚫 Calculators not allowed

1 Find 15% of $260.

_____

2 Solve $5x^2 + 1 = 81$.

_____

3 Find w.

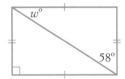

_____

4 For this trapezium, find:

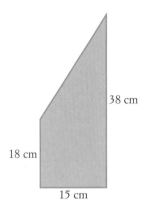

38 cm

18 cm

15 cm

a its perimeter

_____

_____

b its area

_____

5 This stem-and-leaf plot has a missing value.

Stem	Leaf
13	2 3
14	0 5 6 7
15	1 ☐ 6 8 8 9
16	3 4 4 5

a What is the value of ☐ if the median of the data is 154?

_____

_____

b Find the range of the data.

_____

c Describe the shape of the distribution.

_____

PART B: REVIEW

1 For each equation, find the missing power.

a $625 = 5^{☐}$. _____

b $2401 = 7^{☐}$. _____

2 Simplify each expression.

a $3a^4b^2 \times 2a^3b^5$

b $\dfrac{20x^9}{4x}$

c $45m^6n^8 \div (3mn^2)^2$

3 Make k the subject of each formula.

a $y = a + kx$

b $p = \sqrt{\dfrac{k}{m}}$

c $R(1 + k) = 1 - k$

PART C: PRACTICE

 › Logarithms
› Logarithm laws

1 Rewrite each equation in logarithmic form.

a $25^{\frac{1}{2}} = 5$

b $\dfrac{1}{\sqrt{6}} = 6^{-\frac{1}{2}}$

2 Evaluate each expression.

a $\log_3 270 - (\log_3 2 + \log_3 5)$

b $\dfrac{1}{2}\log_4 25 - 2\log_4 \sqrt{20}$

c $\dfrac{1}{3}\log_2 125 - 3\log_2 \sqrt[3]{80}$

3 Simplify each expression.

a $\log_a a^2 + 3\log_a a^3$

b $\dfrac{\log_a(x^7)}{\log_a x}$

c $\log_a \sqrt{x} - \log_a \dfrac{1}{x}$

PART D: NUMERACY AND LITERACY

1 What does the logarithm of a number mean?

2 Complete each logarithm law.

a The logarithm of a product of terms is equal to _____ of the logarithm of each term.

b The logarithm of a quotient of terms is equal to _____ between the logarithm of each term.

c The logarithm of a term raised to a power is equal to the power _____ by the logarithm of the term.

3 If $\log_{10} 3 \approx 0.4771$, find the value of:

a $\log_{10} 9$

b $\log_{10} 300$

c $\log_{10}\left(\dfrac{10}{3}\right)$

d $\log_{10}\left(\sqrt{90}\right)$

(7) LOGARITHMS 2

YOU CAN SOLVE THE EXPONENTIAL EQUATION $3^x = 243$ USING LOGARITHMS.

Name:

Due date:

Parent's signature:

Part A	/ 8 marks
Part B	/ 8 marks
Part C	/ 8 marks
Part D	/ 8 marks
Total	/ 32 marks

PART A: MENTAL MATHS

🚫 Calculators not allowed

1 Test whether each point lies on the line $x - 2y = 5$.

a $(9, -5)$

_____

b $(17, 6)$

_____

2 Increase $7000 by 15%.

_____

3 (3 marks) Simplify each expression.

a $\dfrac{2h^2 + 18h}{-4h - 36}$

b $\dfrac{5}{y^2 - 4} \div \dfrac{15}{2y + 4}$

_____

4 Shoppers at a mall were asked whether they had a pet dog or cat. The results are shown in the two-way-table.

	Cat	No cat
Dog	15	25
No dog	10	30

a How many shoppers had a cat?

_____

b Find the probability of randomly selecting a shopper from the survey who does not have a cat or dog.

_____

5 If $\triangle ABE \mathbin{|||} \triangle ACD$, find d.

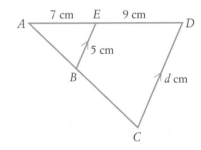

_____

PART B: REVIEW

1 (4 marks) Simplify each expression.

a $\log_x 8 - (\log_x 10 + \log_x 4)$

_____

b $\dfrac{1}{2}(\log_x 8 + \log_x 18)$

_____

2 Simplify each expression.

a $\log_a (a^3)$

_____

b $\log_a y^3 - 3\log_a y$

_____

3 If $\log_{10} 4 = 0.6021$, find the value of:

a $\log_{10} 2.5$

_____

b $\log_{10} \sqrt{40}$

_____

9780170454568

PART C: PRACTICE

> 📝 › Exponential and logarithmic equations

1 Solve each exponential equation.

a $5^x = 78\ 125$

b $6^{n-2} = \dfrac{1}{216\sqrt{6}}$

2 (4 marks) Solve each equation by expressing both sides with a base of 2.

a $4^{2-x} = \dfrac{1}{8}$

b $8^{x+1} = \dfrac{1}{8\sqrt{2}}$

3 Solve each logarithmic equation.

a $\log_x 4 = 2$

b $\log_x 4.8 = \dfrac{1}{2}$

PART D: NUMERACY AND LITERACY

1 Complete: Exponential equations are equations like $3^x = 243$, where the variable is a _____.

2 Solve each equation, correct to 3 decimal places.

a $8^{5-x} = 4000$

b $7^{k+5} = 300$

3 (2 marks) Aveen invests $15 000 at 1% per month compound interest. How many whole months will it take for Aveen's investment to grow to $20 000?

4 (3 marks) A radioactive subtance with a mass of 150 g decays according to the equation

$$A = 150 \times 2^{-0.05t}$$

where A is the amount in grams remaining after t days. Find, correct to the nearest whole number:

a the mass of substance remaining after 10 days.

b the time taken for the substance to decay to a mass of 20 g.

⑦ CHANGING THE SUBJECT OF A FORMULA

CHANGE THE SUBJECT OF EACH FORMULA TO THE VARIABLE INDICATED IN BRACKETS.

1 $I = Prn$ $[r]$

2 $V = lwh$ $[w]$

3 $s = \dfrac{d}{t}$ $[t]$

4 $C = 2\pi r$ $[r]$

5 $l = \dfrac{\theta}{360} \times 2\pi r$ $[r]$

6 $A = \pi r^2$ $[r]$

7 $V = Ah$ $[A]$

8 $A = \dfrac{1}{2}xy$ $[x]$

9 $V = \pi r^2 h$ $[h]$

10 $A = s^2$ $[s]$

11 $A = \dfrac{1}{2}bh$ $[b]$

12 $A = \dfrac{1}{2}(a + b)h$ $[h]$

13 $A = \dfrac{\theta}{360} \times \pi r^2$ $[r]$

14 $V = \pi r^2 h$ $[r]$

15 $A = lw$ $[w]$

16 $y = mx + c$ $[c]$

17 $C = \pi d$ $[d]$

18 $A = P(1 + r)^n$ $[P]$

19 $V = \dfrac{4}{3}\pi r^3$ $[r]$

20 $A = 180(n - 2)$ $[n]$

21 $y = mx + c$ $[m]$

22 $SA = 4\pi r^2$ $[r]$

23 $ax + by + c = 0$ $[x]$

24 $ax + by + c = 0$ $[y]$

25 $A = \dfrac{1}{2}(a + b)h$ $[a]$

26 $c^2 = a^2 + b^2$ $[b]$

27 $SA = \pi r^2 + \pi rl$ $[l]$

28 $y - y_1 = m(x - x_1)$ $[x]$

29 $SA = 2\pi rh + 2\pi r^2$ $[h]$

30 $A = P(1 + r)^n$ $[r]$

STARTUP ASSIGNMENT 8 (8)

THIS ASSIGNMENT REVISES WORK ON
ALGEBRA AND GRAPHING LINES TO PREPARE
YOU FOR THE GRAPHING CURVES TOPIC.

PART A: BASIC SKILLS   / 15 marks

1 Evaluate $\dfrac{12}{\sqrt{5^2 + 7^2}}$, correct to 2 decimal places.

_____

2 Simplify $4^{\frac{1}{2}} \times 4^{-2}$. _____

3 Complete: 4 m² = _____ cm²

4 Complete: 55 m/s = _____ km/h

5 Find the value of d in the diagram below, correct to 2 decimal places.

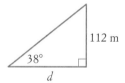

_____

6 Simplify $\dfrac{20x^3 y^2}{4x^3 y}$. _____

7 Find, correct to 3 significant figures, the volume of a cylinder of radius 10 cm and height 12 cm.

_____

8 Evaluate $\dfrac{2.46 \times 10^{12}}{4 \times 10^8}$. _____

9 What is the angle sum of a parallelogram?

_____

10 If $\dfrac{5}{6}$ of a number is 90, what is the number?

_____

_____

11 A number is chosen at random from the numbers 1 to 20. What is the probability that it is a multiple of 3?

_____

12 Find θ, to the nearest degree, if cos θ = 0.561.

_____

13 Divide $4550 in the ratio 2 . 1 . 4. _____

_____

_____

14 How many axes of symmetry has a parallelogram? _____

15 Madeline is paid $275 plus a 5.5% commission on items sold. Calculate her pay when she sells $3300 worth of items.

_____

PART B: ALGEBRA AND GRAPHS   / 25 marks

16 a Complete the table below for $y = 3x - 2$.

x	-2	-1	0	1	2	3
y		-5			4	

b Graph $y = 3x - 2$ on a number plane.

c What is the gradient? _____

d What is the y-intercept? _____

e What is the x-intercept? _____

17 Complete the table below for $y = x^2 - 4$.

x	-2	-1	0	1	2	3
y		-3			0	

18 Solve each equation.

a $3x + 7 = 1$ _____

_____

b $\dfrac{2y + 4}{5} = 8$ _____

_____

19 Draw a line with a positive gradient.

20 Evaluate $3x^2$ if:

a $x = 2$ _____

b $x = -4$ _____

c $x = -1$ _____

21 Write the formula for each table of values.

a

x	-2	-1	0	1	2	3
y	-5	-1	3	7	11	15

_____

b

x	-2	-1	0	1	2	3
y	4	1	0	1	4	9

_____

22 This graph shows Debbie's day trip.

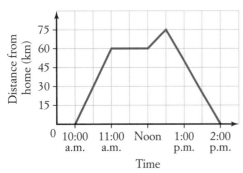

a What was Debbie's speed during the first hour? _____

b When did she begin her trip home?

_____

c What total distance did she travel?

_____

23 For this line, find:

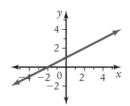

a the y-intercept _____

b the gradient _____

c its equation. _____

24 Evaluate $\dfrac{4}{x}$ if:

a $x = -3$ _____

b $x = 10$ _____

c $x = \dfrac{1}{2}$ _____

PART C: CHALLENGE   Bonus   / 3 marks

A lady-bug walks along the edges of this prism, starting and ending its journey at vertex A.

What is the maximum distance it can walk without travelling along the same edge twice?

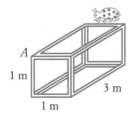

_____

MATCH EACH EQUATION
WITH ITS CORRECT GRAPH.

1 $y = x^2 + 2$

2 $y = (x - 2)^2$

3 $y = -3(x^2 - 2)$

4 $y = (x + 2)^2 + 3$

5 $y = (x - 2)^2 - 3$

6 $y = 3(x + 2)^2 + 4$

7 $y = x^2 - 2$

8 $y = 3(x - 2)^2 + 4$

9 $y = (x + 2)^2$

10 $y = (x + 2)^2 - 3$

11 $y = 3(x^2 - 2)$

12 $y = 3(x - 2)^2 - 4$

13 $y = 3(x + 2)^2$

14 $y = -3(x + 2)^2 + 4$

15 $y = (x - 2)^2 + 3$

16 $y = -3(x + 2)^2$

A

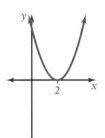

B

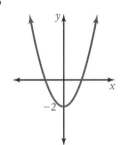

C

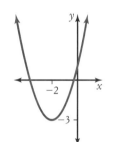

D

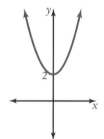

E

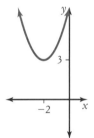

F

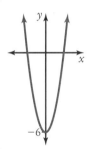

G

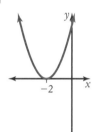

H

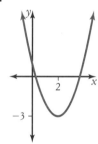

I

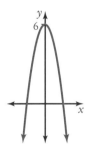

J

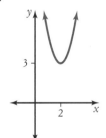

K

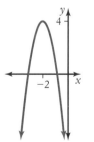

L

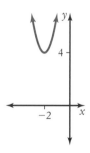

M

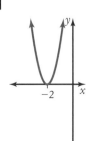

N

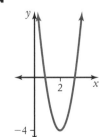

O

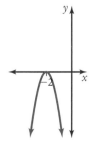

P

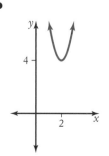

Chapter 8 Graphing curves **93**

8 GRAPHING EXPONENTIALS

SOME GRAPH PAPER TO HELP
YOU GRAPH SOME CURVES.

Teacher's tickbox

Graph the ticked set of exponential equations.

❏ $y = 2^x, y = 2^{-x}, y = -2^x, y = -2^{-x}$ ❏ $y = 3^x, y = 3^{-x}, y = -3^x, y = -3^{-x}$

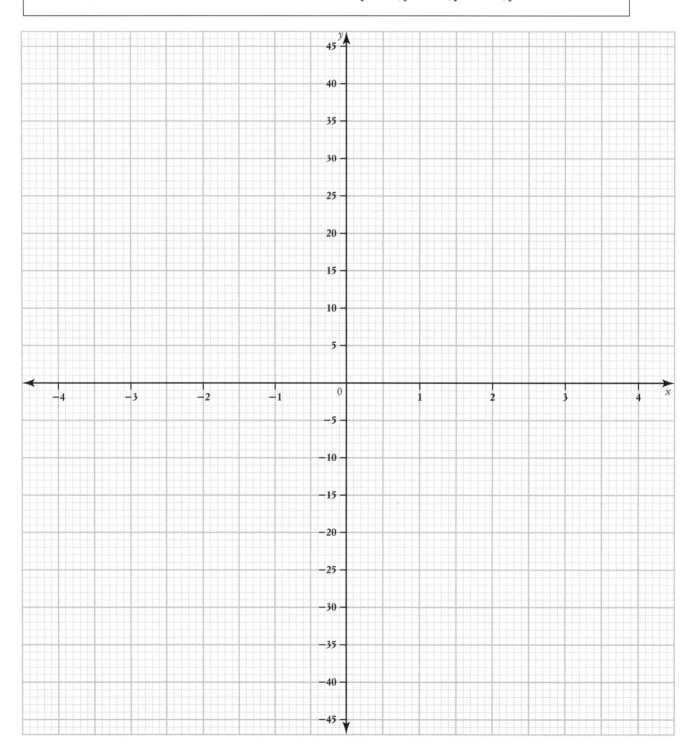

GRAPHING HYPERBOLAS (8)

THESE GRAPHS HAVE ASYMPTOTES THAT YOU SHOULD SHOW AS DOTTED LINES.

Teacher's tickbox

Graph the ticked pair of equations.

❏ $y - \dfrac{2}{x}, y - \dfrac{1}{x}$ ❏ $y = \dfrac{2}{x} + 1, y = \dfrac{3}{x} - 2$ ❏ $y - \dfrac{3}{x}, y - -\dfrac{5}{x}$ ❏ $y = \dfrac{3}{x-1}, y = -\dfrac{2}{x+2}$

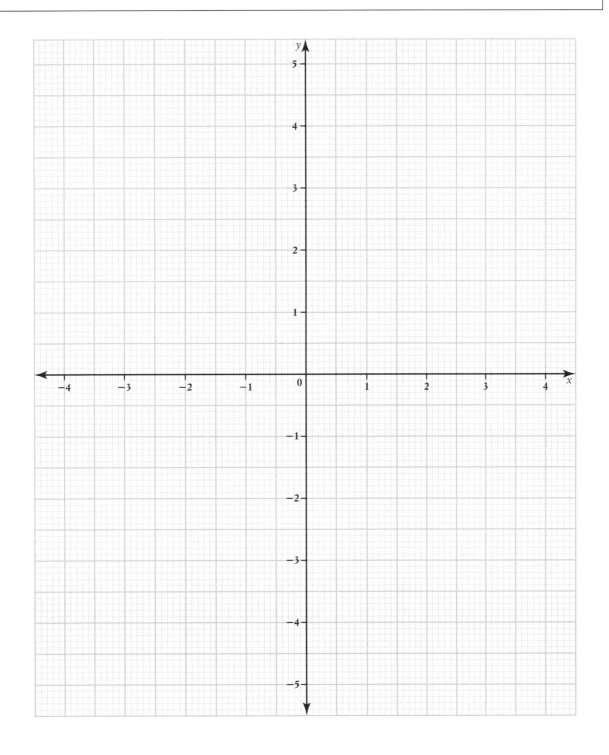

GRAPHING CURVES CROSSWORD

SOME WORDS FROM THIS TOPIC
ARE SHOWN NEXT PAGE, BUT
THE LETTERS ARE JUMBLED.

Unscramble the word clues and place them in the crossword.

9780170454568

Clues across

3 NEIL

6 TAMYESOPT

7 PHRAG

9 PINOT

10 CLERIC

11 RECENT

13 CREDIT

18 DECORATIONS

19 VEXTER

22 CANSNOTT

24 VERBIALA

25 HAZITLONOR

26 SIXA

Clues down

1 AUTOENQI

2 EVOCCAN

4 NIXANTELOPE

5 CARATQUID

8 CENTIOFFICE

11 VURCE

12 SNORNOVICE

14 NEVERIS

15 TALCRIVE

16 RIGION

17 PROIPROTON

20 SURADI

21 CENTREPIT

23 ALBOARPA

8 GRAPHING CURVES

Name:

Due date:

Parent's signature:

Part A	/ 8 marks
Part B	/ 8 marks
Part C	/ 8 marks
Part D	/ 8 marks
Total	/ 32 marks

HI, MITCH HERE. THIS ASSIGNMENT REVISES ALGEBRA AND GRAPHING SKILLS. KEEP PRACTISING UNTIL IT ALL MAKES SENSE.

PART A: MENTAL MATHS

Calculators not allowed

1 Convert $\dfrac{25}{40}$ to a percentage.

_____

2 Simplify each expression.

a $(x + 3)(x - 4)$

_____

b $\dfrac{a^3b^2}{3ab}$ _____

c $(6m)^{-2}$ _____

3 Name the 2 quadrilaterals that have all sides equal.

_____

_____

4 A jar contains only blue and white marbles. The probability of selecting a blue marble from the jar is 20%.

a What is the probability of selecting a white marble? _____

b If there are 25 marbles in the bag, how many blue marbles are there?

_____

PART B: REVIEW

1 a If R is directly proportional to N and 0.74 is the constant of proportionality, write an equation for R.

_____

b Find the value of R when $N = 5$.

_____

c If this equation was graphed on a number plane, what type of graph would it be and what would be its y-intercept?

_____

_____

2 (2 marks) T varies directly with s. When $s = 5$, $T = 203.5$. Find T when $s = 24.1$.

_____

_____

_____

3 (3 marks) Find the value of y when $x = 0$ in each equation.

a $y = 5x^2 + 8$

_____

b $y = 5^x$

_____

c $x^2 + y^2 = 16$

_____

_____

PART C: PRACTICE

> The parabola $y = ax^2 + c$
> The exponential curve $y = a^x$
> The circle $x^2 + y^2 = r^2$

1 (3 marks) Match each quadratic equation to its graph.

a $y = x^2$ **b** $y = -x^2 + 3$ **c** $y = x^2 - 1$

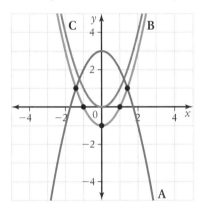

2 (4 marks) Match each exponential equation to its graph.

a $y = 2^x$ **b** $y = 2^{-x}$

c $y = -2^x$ **d** $y = -2^{-x}$

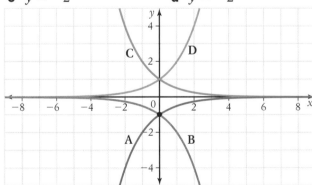

3 Write the equation of a circle with centre (0, 0) and radius 4.

_____

PART D: NUMERACY AND LITERACY

1 a Is the graph of $y = -3x^2 + 1$ concave up or down? Give reasons.

_____

_____

b How many axes of symmetry does this graph have?

_____

2 (2 marks) Graph $x^2 + y^2 = 9$ and state its centre and radius.

3 (2 marks) Sketch the graph of $y = 5^x - 1$ and state its asymptote.

4 (2 marks) Graph $y = \dfrac{1}{2}x^2 + 1$ and write the coordinates of its vertex.

⑧ GRAPHING CURVES 2

SOME QUITE ADVANCED GRAPHS HERE: PARABOLAS, EXPONENTIALS, HYPERBOLAS, CIRCLES.

Name:

Due date:

Parent's signature:

Part A	/ 8 marks
Part B	/ 8 marks
Part C	/ 8 marks
Part D	/ 8 marks
Total	/ 32 marks

PART A: MENTAL MATHS

🔲 Calculators not allowed

1 (3 marks) Prove that $\triangle RST \;|||\; \triangle UVW$, then find n.

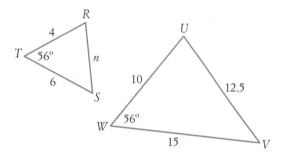

_____

_____

_____

_____

_____

_____

_____

2 Simplify $\sqrt{20} + \sqrt{45}$.

_____

_____

3 (2 marks) Factorise $-6x^2 - 27x + 15$.

_____

_____

_____

4 Convert each statement to an algebraic expression.

a How many 5s divide into x

_____

b 6 less than y, all divided by 7

_____

PART B: REVIEW

1 a Find the x- and y-intercepts of the graph of $y = 3(x + 4)^2$. _____

_____

_____

_____

b Sketch the graph of $y = 3(x + 4)^2$.

9780170454568

2 Match each graph with its correct equation.

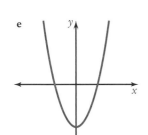

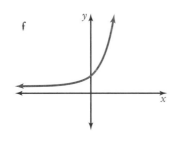

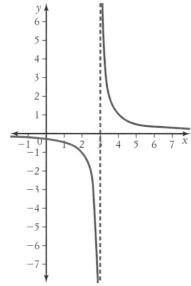

A $y = x^2$

B $y = 2^x$

C $y = \dfrac{1}{2}x^2$

D $y = x - 1$

E $y = x^2 - 4$

F $y = 4x^2$

PART C: PRACTICE

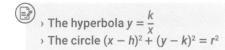

› The hyperbola $y = \dfrac{k}{x}$

› The circle $(x - h)^2 + (y - k)^2 = r^2$

1 Match each graph with its correct equation.

a

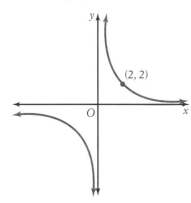

(2, 2)

c

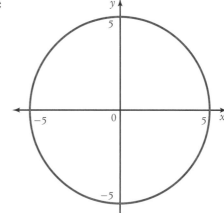

d

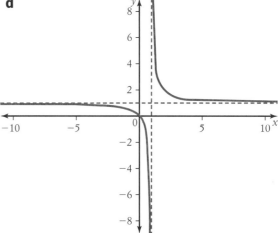

e

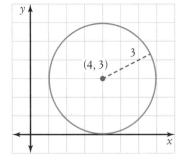

f

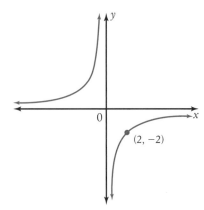

A $y = \dfrac{1}{x-1} + 1$

B $x^2 + y^2 = 25$

C $y = \dfrac{4}{x}$

D $(x-4)^2 + (y-3)^2 = 9$

E $y = \dfrac{1}{x-3}$

F $y = -\dfrac{4}{x}$

2 (2 marks) Find the centre and radius of the circle with equation

$x^2 - 12x + y^2 + 4y + 24 = 0$.

_____

_____

_____

_____

_____

_____

1 (3 marks) Complete:

a The hyperbola $y = -\dfrac{3}{x}$ has branches in the

_____ and _____ quadrants.

b The x- and y-axes are called _____ because

the hyperbola $y = -\dfrac{3}{x}$ approaches them but

never crosses them.

c The hyperbola $y = -\dfrac{3}{x} - 5$ is the hyperbola

$y = -\dfrac{3}{x}$ moved _____ 5 units.

2 (5 marks) Sketch the graph of each equation, showing all intercepts and asymptotes.

a $y = -\dfrac{1}{x-4}$

b $y = -\dfrac{2}{x} - 3$

c $x^2 + (y+2)^2 = 9$

THIS ADVANCED TOPIC IS
FURTHER TRIGONOMETRY.

PART A: BASIC SKILLS   / 15 marks

1 35% of an amount is $87.50. What is the
amount? _____

2 Evaluate $16^{\frac{3}{4}}$. _____

3 How many hours and minutes will it take a car
travelling at 72 km/h to cover 120 km?

_____

_____

4 Convert $0.7\dot{2}\dot{}$ to a simple fraction.

_____

_____

_____

5 Find the perimeter of this trapezium.

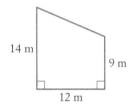

14 m

9 m

12 m

_____

_____

_____

6 Expand and simplify $(3+\sqrt{5})^2$.

_____

_____

7 If a cube has volume 195.112 cm³, find its
surface area. _____

8 Calculate to the nearest cent the compound
interest from $16 000 invested at 6.2% p.a. for
8 years.

_____

_____

9 Find the gradient of the line perpendicular to
$4x - 3y + 1 = 0$.

_____

_____

10 Factorise $4y^2 - 17y - 15$.

_____

_____

11 Find d.

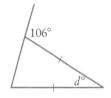

106°

$d°$

12 Solve $5 + 2k \le 6 - k$. _____

_____

_____

13 Find the interquartile range of these numbers:
16, 17, 24, 10, 16, 14, 13, 14, 19

_____

14 Solve $-5y^3 = 135$.

_____

_____

C
S
F

_____

15 Find the equation of the line passing through $(0, 0)$ and $(3, -1)$.

_____

_____

PART B: TRIGONOMETRY / 25 marks

16 Write $\cos \theta$ as a fraction. _____

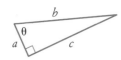

17 Calculate to 4 decimal places:

a $\dfrac{12 \sin 70°}{10}$ _____

b $\dfrac{14 \sin 16°}{\sin 42°}$ _____

c $\dfrac{3^2 + 5^2 - 2^2}{2 \times 3 \times 5}$ _____

d $\cos 70°$ _____

e $\sin 20°$ _____

18 a What do you notice about $\cos 70°$ and $\sin 20°$ in the question above?

_____

b $\cos$ _____ $= \sin 50°$

19 Using the triangle, write as a fraction:

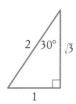

d $\cos 30°$ _____

e $\tan 30°$ _____

20 Find θ correct to the nearest minute.

a $\tan \theta = \dfrac{3}{5}$ _____

b $\cos \theta = 0.553$ _____

c $\sin \theta = \dfrac{\sqrt{3}}{2}$ _____

d

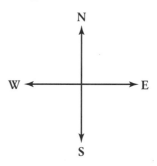

_____

21 Draw a bearing of 200°.

```
        N
        ↑
        |
W ←-----+-----→ E
        |
        ↓
        S
```

22 If $a = 6$, $b = 8$ and $C = 40°$, evaluate to 2 decimal places:

a $\dfrac{1}{2} ab \sin C$ _____

b $a^2 + b^2 - 2ab \cos C$ _____

23 Find x correct to 2 decimal places.

a $x = \dfrac{25}{\tan 16° \, 10'}$ _____

b $x^2 = 3^2 + 9^2 - 2 \times 3 \times 9 \cos 27°$ _____

_____

c

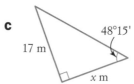

_____

d

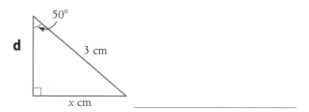

_____

24 The angle of elevation from a point 550 m from the base of a tower to the top is 42° 25′. Calculate the height of the tower to 2 decimal places.

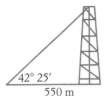

_____

_____

25 Find (to the nearest degree) the angle of depression of a boat that is 200 m from the base of a 74 m cliff.

_____

_____

26 From the lookout, a hiker walks north 8 km, then 6 km west to the camp.

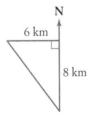

a How far is the camp from the lookout?

_____

b What is the bearing (in degrees and minutes) of the camp from the lookout?

_____

PART C: *CHALLENGE*   Bonus   / 3 marks

Find an angle size θ that satisfies the equation sin θ = cos θ. Can you prove your answer?

⑨ FINDING AN UNKNOWN ANGLE

YOU NEED TO USE THE SINE OR
COSINE RULES TO SOLVE THESE

Find the size of the angle marked in each triangle, correct to the nearest degree.

1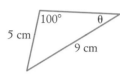

2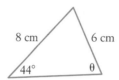

3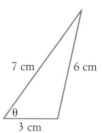

4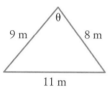

5 θ is obtuse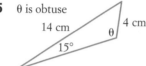

6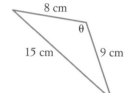

7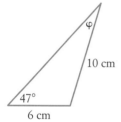

8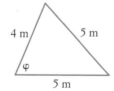

9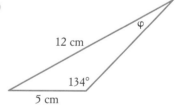

10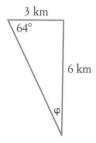

11 φ is obtuse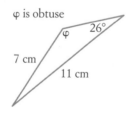

12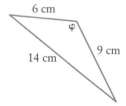

13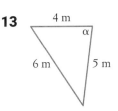

4 m
α
6 m
5 m

14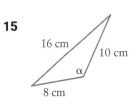

5 m
α
13 m
20°

15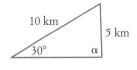

16 cm
10 cm
α
8 cm

16 α is obtuse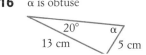

20°
α
13 cm
5 cm

17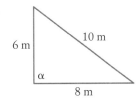

6 m
10 m
α
8 m

18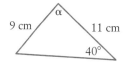

10 km
5 km
30°
α

19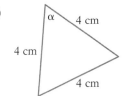

α
4 cm
4 cm
4 cm

20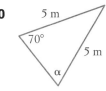

5 m
70°
5 m
α

21

9 cm
α
11 cm
40°

Mixed answers: 83°, 68°, 136°, 90°, 80°, 70°, 125°, 63°, 60°, 66°, 88°, 117°, 33°, 115°, 58°, 27°, 52°, 26°, 137°, 124°, 17°, 90°

9 TRIGONOMETRY CROSSWORD (ADVANCED)

HOW WELL DO YOU KNOW YOUR ADVANCED TRIGONOMETRY TERMINOLOGY?

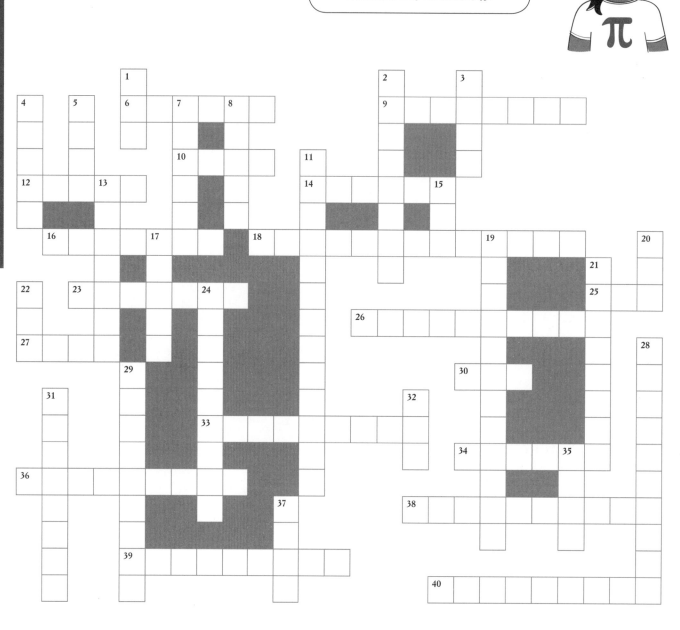

Across

6 $\dfrac{1}{3600}$ of a degree

9 The sign of tan of an obtuse angle

10 cos 45° is an exact ratio because its value is a _____.

12 The Greek letter θ

14 An angle between 90° and 180°

16 There are 60 of them in a degree.

18 Angles that add to 180°

23 An angle that shows direction

25 tan 45°

26 The cosine rule is an extension of his theorem.

27 Where does the Sun set?

30 The compass direction with a bearing of 292.5°

33 Angle of 'looking up'

34 A unit for measuring angles

36 The direction of Darwin from Sydney

38 Opposite of **33** across

39 The opposite of north-east

40 _____ angles between parallel lines are equal.

Down

1 Halfway between south-east and east

2 The angle between 2 sides in a triangle

3 cos 60°

4 The direction of Antarctica from anywhere

5 This rule involves side–angle pairs in a triangle.

7 Complementary to sine

8 On a compass rose, the direction pointing upwards

11 Angles in a right angle are this.

13 The full name of tan

15 The compass direction with bearing 067.5°

17 tan 60° is the square root of this.

19 The mathematics of triangle measurement

20 The opposite of NNW

21 The sign of sine of an obtuse angle

22 The direction 22.5° west of north

24 The direction of Sydney from Canberra

28 The longest side of a right-angled triangle

29 SE stands for this direction.

31 The sine rule relates a triangle's sides to their _____ angles.

32 The direction 45° east of NNW

35 The opposite of **27** across

37 What A represents in the formula $A = \dfrac{1}{2}\,ab \sin C$

9 FURTHER TRIGONOMETRY 1

HERE WE'LL LOOK AT THE TRIGONOMETRY OF ANGLES GREATER THAN 90°.

Name:

Due date:

Parent's signature:

Part A	/ 8 marks
Part B	/ 8 marks
Part C	/ 8 marks
Part D	/ 8 marks
Total	/ 32 marks

PART A: MENTAL MATHS

🚫 Calculators not allowed

1 Simplify $4\sqrt{18} + \sqrt{98}$.

_____

_____

2 Mai earns a salary of $78 000 p.a. How much does she earn each month?

_____

_____

3 Find the gradient of the line passing through points (4, 5) and (–6, 7).

_____

_____

4 Simplify.

a $\left(\dfrac{2m^4 n^5}{m^5 n^3}\right)^{-2}$

_____

_____

b $\left(9x^2\right)^{-\frac{3}{2}}$

_____

_____

5 If 3 coins are tossed together, what is the probability of tossing 1 head and 2 tails?

_____

_____

6 (2 marks) Solve $2x^2 - 9x - 5 = 0$.

_____

_____

_____

PART B: REVIEW

1 For this rectangular prism, find:

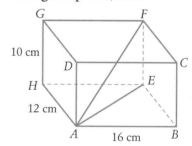

a AE

_____

_____

b AF, correct to one decimal place.

_____

_____

9780170454568

c ∠*FAE*, correct to the nearest degree.

_____

_____

b the size of angle *A*.

_____

_____

2 The bearing of Braddon (*B*) 182 km from Ashville (*A*) is 138°.

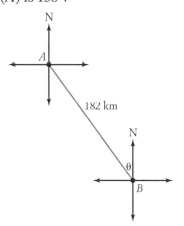

a Find the value of θ.

_____

b How far south of Ashville is Braddon, to the nearest kilometre?

_____

_____

c What is the bearing of Ashville from Braddon?

_____

_____

3 Find:

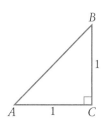

a the length of *AB* as a surd

_____

_____

PART C: PRACTICE

> › Trigonometry of angles greater than 90°
> › Trigonometric equations

1 Evaluate tan 158° 22' correct to 2 decimal places.

_____

2 If θ is acute, find θ if:

a sin 135° = sin θ _____

b cos 100° = −cos θ _____

c tan 153° = −tan θ _____

3 (2 marks) Solve sin θ = 0.74 correct to the nearest degree, giving all possible acute and obtuse solutions.

_____

_____

_____

_____

4 (2 marks) Solve $\cos x = -\dfrac{3}{25}$, correct to the nearest minute, if *x* is obtuse.

_____

_____

_____

PART D: NUMERACY AND LITERACY

1 This is the graph of $y = \cos \theta$.

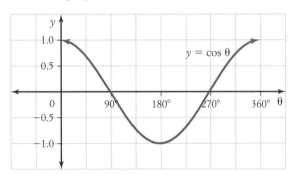

a If θ is obtuse, is $\cos \theta$ positive or negative?

_____

b What is $\cos 90°$?

_____

c What is the highest value of $y = \cos \theta$?

_____

d For what values of θ is $\cos \theta = 0.5$?

_____

2 $\triangle CBA$ is an equilateral triangle.

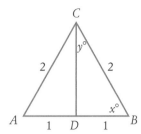

a Find its height CD as a surd.

_____

_____

b Find the values of x and y.

_____

c Hence find the value of $\sin 60°$ as a surd.

_____

9780170454568

Name:

Due date:

Parent's signature:

Part A	/ 8 marks
Part B	/ 8 marks
Part C	/ 8 marks
Part D	/ 8 marks
Total	/ 32 marks

THE SINE AND COSINE RULES
WORK FOR TRIANGLES THAT
ARE NOT RIGHT-ANGLED.

PART A: *MENTAL MATHS*

🚫 Calculators not allowed

1 Evaluate $125^{-\frac{2}{3}}$.

_____

2 Rationalise the denominator of $\dfrac{18}{2\sqrt{2}}$.

_____

_____

3 (2 marks) Factorise $10\,000b^4 - 81a^4$.

_____

_____

4 (2 marks) Find the equation of the line perpendicular to $y = 5x - 2$ that passes through the point $(1, 4)$.

_____

_____

_____

5 For the graph of $y = 2x^2 - 8$, find:

a the y-intercept _____

_____

b the x-intercepts _____

_____

PART B: *REVIEW*

1 (3 marks) From the top of a 200 m tower, the angle of depression of a car is 50°.

a Draw a diagram to show this information.

b How far, correct to the nearest metre, is the car from the foot of the tower?

_____

_____

2 Which compass direction is:

a 270°? _____

b 135°? _____

3 (3 marks) A ship is due south of Adelaide. From the ship, on a bearing of 285°, a lighthouse is seen. If the lighthouse is 12 km due west of Adelaide, how far is the ship from the lighthouse, to one decimal place?

_____

_____

_____

HOMEWORK

HW

9780170454568

PART C: PRACTICE

› The sine and cosine rules

1 (4 marks) Find the value of each variable, correct to 2 decimal places.

a

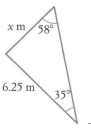

_____

_____

b

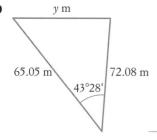

_____

_____

_____

2 (4 marks) Find the value of each variable, correct to the nearest minute.

a

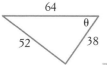

_____

_____

b

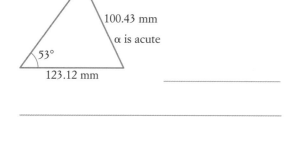

_____

_____

PART D: NUMERACY AND LITERACY

1 Which rule is a relationship between the 3 sides of a triangle and one of its angles?

_____

2 Rewrite the cosine rule $c^2 = a^2 + b^2 - 2ab \cos C$ so that $\cos C$ is the subject of the formula.

_____

_____

_____

3 Complete: In any triangle, the ratios of the sides to the sine of their _____ angles are _____ .

4 (2 marks) Find θ to the nearest degree if it is obtuse.

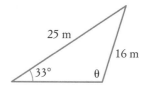

_____

_____

_____

5 (3 marks) A plane flew on a bearing of 135° for 275 km. It then changed direction and flew another 316 km on a bearing of 223°. How far, correct to the nearest kilometre, is the plane from its starting point?

_____

_____

_____

_____

Name:

Due date:

Parent's signature:

Part A	/ 8 marks
Part B	/ 8 marks
Part C	/ 8 marks
Part D	/ 8 marks
Total	/ 32 marks

THE FORMULA IN PART C IS FOR THE AREA OF A TRIANGLE WHEN YOU KNOW THE LENGTHS OF 2 SIDES AND THE INCLUDED ANGLE BETWEEN THEM.

PART A: *MENTAL MATHS*

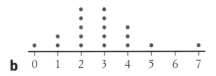 Calculators not allowed

1 Find the interquartile range of each set of data.

a 4　7　7　9　11　12　13　14　17

_____

```
        •   •
        •   •   •
    •   •   •   •
•   •   •   •   •   •       •
0   1   2   3   4   5   6   7
```

b 0　1　2　3　4　5　6　7

_____

2 (2 marks) Simplify $\dfrac{x^2 - 4}{2x^2 - 9x + 10}$.

_____

_____

_____

_____

3 (2 marks) Find, in terms of π, the surface area of a cylinder with base radius 5 m and height 10 m.

_____

_____

_____

_____

4 (2 marks) Sketch the graph of $y = 2^x$, showing the y-intercept and the coordinates of one point.

PART B: *REVIEW*

1 (2 marks) Find the value of x, correct to the nearest whole number.

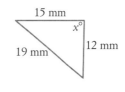

15 mm

$x°$

19 mm

12 mm

_____

_____

_____

_____

2 (4 marks) The angles of elevation of a building measured from 2 positions 30 m apart are 42° and 60°.

a Find ∠ADB.

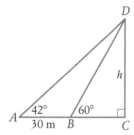

_____

b Find, correct to 2 decimal places, the length of BD.

_____

_____

_____

c Hence find the height, CD, of the building, correct to the nearest metre.

_____

_____

3 (2 marks) Find x, correct to one decimal place.

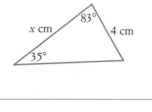

_____

_____

_____

PART C: PRACTICE

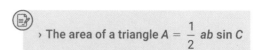

› The area of a triangle $A = \dfrac{1}{2} ab \sin C$

1 (5 marks) Find, correct to one decimal place, the area of each shape.

a

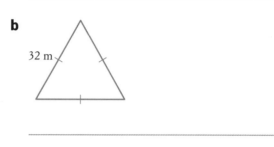

_____

_____

b

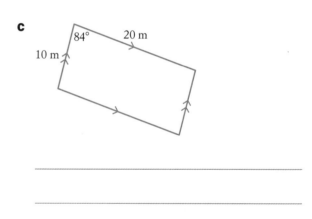

_____

_____

c

_____

_____

2 O is the centre of a circle of radius 35 cm.

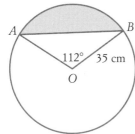

Calculate, correct to one decimal place, the area of:

a sector OAB with angle $112°$

_____

b △OAB

_____

c the shaded segment

_____

PART D: NUMERACY AND LITERACY

1 Complete: The area of a triangle with sides of length a and b and _____ angle C is $A = \dfrac{1}{2}\ ab$ _____.

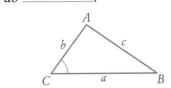

2 Which rule is used for triangle problems involving 2 sides and the 2 angles opposite them?

_____

3 What is another name for the cosine rule if the angle used is $90°$?

_____

4 (4 marks) Ronan wants to run around this cross-country course.

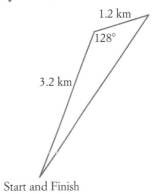

Start and Finish

Calculate, correct to one decimal place:

a the area of the triangle

_____

b the perimeter of the triangle

_____

_____

_____

⑨ THE SINE AND COSINE CURVES

Teacher's tickbox
Graph:
❏ $y = \sin x$
❏ $y = \cos x$

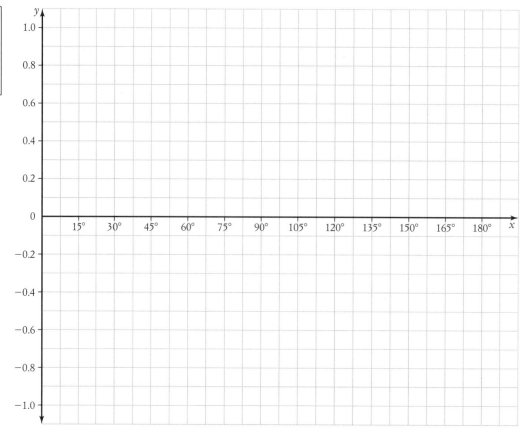

1 Complete the table (to 2 decimal places) for the sine or cosine curve (ticked above).

x	0°	15°	30°	45°	60°	75°	90°	105°	120°	135°	150°	165°	180°
y													

2 Plot the points on the grid above and draw a smooth curve through them.

3 Describe the shape of your curve.

_____

4 Is the curve symmetrical? How? Note also the symmetry in your table of values.

_____

5 What is:

a the maximum value of y? _____

b the minimum value of y? _____

6 For what value(s) of x is it true that:

a $y = 0$? _____ **b** $y = 1$? _____

c $y = -1$? _____ **d** $y = \frac{1}{2}$? _____

e y is positive? _____ **f** y is negative? _____

7 What do you think will happen to the curve when $x > 180°$? _____

Further investigations

1 How are the sine and cosine curves similar?

_____

2 Notice how $\sin 75° = \sin 105°$ and $\sin 75° = \cos 15°$. Investigate whether the following properties are true or false.

a $\sin x = \sin (180° - x)$ _____

b $\cos x = \cos (180° - x)$ _____

c $\sin x = \cos (90° - x)$ _____

d $\cos x = \sin (90° - x)$ _____

3 Graph the sine and cosine curves for $x = 0°$ to $360°$.

4 Graph $y = \tan x$.

LET'S GET READY FOR SIMULTANEOUS EQUATIONS, A NEW TOPIC WHERE YOU SOLVE 2 EQUATIONS TOGETHER.

PART A: BASIC SKILLS   / 15 marks

1 Evaluate $\sqrt{150}$ correct to 3 significant figures.

_____

2 Calculate Arden's fortnightly pay

if he earns a salary of $85 300. _____

3 Write an expression for the next

even number after n if n is odd. _____

4 What is this formula used for?

$A = \dfrac{1}{2}xy$ _____

5 Prove that this triangle is right-angled.

6 Evaluate:

  a $4^3 \times 5^0$ _____

  b 3^{-2} _____

7 Simplify:

  a $4(x + 3) - 2(x - 3)$ _____

  b $\dfrac{2p}{3} \times \dfrac{3p}{10}$ _____

8 Find the area of this sector, correct to

2 decimal places.

9 Express 2.8×10^{-5} in decimal form.

_____

Solve $\dfrac{2x + 7}{3} = 4.$ _____

_____

11 List the 4 tests for congruent triangles.

_____

12 The mean of 13, 19, 12, x and 17 is 16. Find x.

_____

_____

13 Find the curved surface area of this cylinder,

correct to 2 decimal places.

_____

PART B: ALGEBRA AND EQUATIONS   / 25 marks

14 Simplify each expression.

  a $-8(5g - 4)$ _____

  b $3(2y - 10) + 7y$ _____

  c $-4(8 + 5a) + 2(a + 13)$ _____

_____

  d $9(7b - 3) - 4(10 - 3b)$ _____

_____

15 Solve each equation.

  a $3p = -15$ _____

  b $2f + 13 = 30$ _____

  c $14 - 4d = 20$ _____

  d $2.5g - 16 = 32$ _____

e $\dfrac{x + 5}{3} = 6$ _____

f $4a - 3 = a + 8$ _____

_____

16 Find y when $x = 2$.

a $y = 7x + 15$ _____

b $y = 4(17 - 3x)$ _____

c $5x + 2y = 24$ _____

17 Find a when $b = -5$.

a $14 + 8b = a$ _____

_____

b $a = 45 - 9b$ _____

_____

c $5(b + 12) = a$ _____

_____

18 Test whether $x = 3$ is a solution to:

a $50 - 11x = 17$ _____

_____

b $2x + 2 = 20 - 4x$ _____

_____

c $6(x - 5) = 12$ _____

_____

19 Graph each line on the same number plane.

a $y = 4$ **b** $x = 5$

c $y = 2x - 3$ **d** $y = -x + 5$

20 Use an equation to find 2 consecutive even numbers that add to 150.

_____

_____

_____

_____

21 Use an equation to find 3 consecutive odd numbers that add to 105.

_____

_____

_____

_____

PART C: CHALLENGE   Bonus   / 3 marks

Robbie Rabbit and Timmy Turtle had a race. Robbie ran at a speed of 10 km/h for half the distance, then 8 km/h for the other half. Timmy ran 9 km/h for the entire distance. Who won the race, Robbie or Timmy?

_____

_____

_____

_____

INTERSECTION OF LINES ⑩

IN THIS ACTIVITY, YOU WILL GRAPH
2 LINES AND SEE WHERE THEY CROSS.

Graph and label each pair of linear functions on the number plane below. It may help to plot each pair of lines with a different colour. Write the point of intersection of each pair of lines.

1 $y = x + 2$

$y = 3x - 4$

(___, ___)

2 $y = 2x - 2$

$y = -x + 1$

(___, ___)

3 $y = -x + 2$

$y = -\frac{1}{4}x - 1$

(___, ___)

4 $y = \frac{1}{2}x + 7$

$y = -3x$

(___, ___)

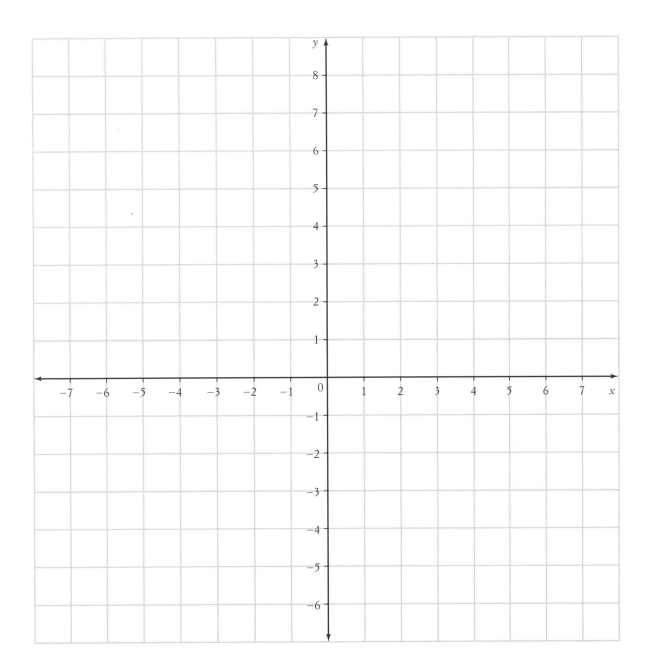

UNSCRAMBLE THE KEYWORDS BELOW AND PLACE THEM IN THE CROSSWORD.

Clues across

7 LOVES

11 NORENTCITIES

12 OPTIN

14 STAYSIF

15 IBISSTUNTOUT

16 EDMOTH

17 OUSTLION

Clues down

1 EMAILSUNTOUS

2 LOBPERM

3 BRAVELAI

4 ECLAIRBAG

5 EXSA

6 HAPGARLIC

8 SOANTIQUE

9 IMINITALONE

10 ENLIAR

13 NIECETICOFF

Name:

Due date:

Parent's signature:

Part A	/ 8 marks
Part B	/ 8 marks
Part C	/ 8 marks
Part D	/ 8 marks
Total	/ 32 marks

SIMULTANEOUS EQUATIONS
CAN BE SOLVED GRAPHICALLY
OR ALGEBRAICALLY.

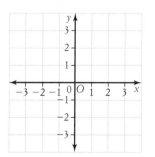

PART A: **MENTAL MATHS**

Calculators not allowed

1 Evaluate $\sqrt{225}$. _____

2 Classify each type of data as categorical or numerical. If numerical, then classify as discrete or continuous.

a The number of people at a meeting.

_____

b The amount of rain in Canberra today.

_____

c The suburb of your home.

_____

3 Find the area of each shape as a simplified algebraic expression.

a

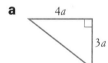

4a

3a _____

_____

b

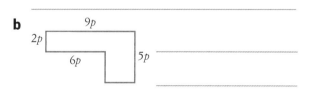

9p

2p

6p   5p _____

_____

_____

4 Find the mode of these data values.

Score	Frequency
13	4
14	6
15	7
16	8

_____

5 Simplify $(2x^2y^3)^3$. _____

PART B: **REVIEW**

1 For $y = 5x + 3$, find y when:

a $x = -2$

_____

b $x = -0.1$

_____

2 (2 marks) Complete the table of values for $x - y = 2$, then graph the equation on the number plane.

x	−1	0	1	2
y				

HOMEWORK

HW

3 Test whether the point $(-1, -1)$ lies on the line:

a $y = 3 - 2x$

_____

b $6x - y = -5$

_____

4 Write the gradient and y-intercept of each linear equation, then graph the equation.

a $y = \dfrac{3}{5}x - 3$

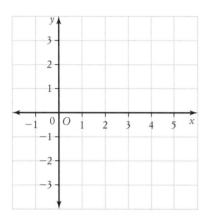

b $y = -\dfrac{1}{2}x + 3$

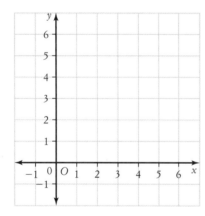

PART C: PRACTICE

› Graphical solution
› The elimination method

1 (4 marks) Complete the table of values for both equations, then solve the 2 simultaneous equations graphically.

$2x + y = 3$

x	0	1	2	3	4
y					

$x + y = -1$

x	0	1	2	3	4
y					

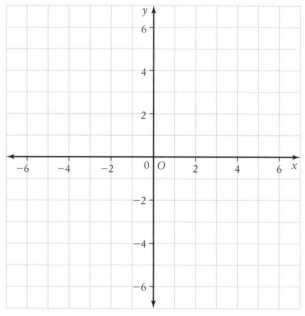

$x =$ _____, $y =$ _____

2 (4 marks) Solve these simultaneous equations using the elimination method.

$$8m - 5n = -21$$
$$4m + 5n = 45$$

_____

_____

_____

_____

_____

PART D: NUMERACY AND LITERACY

1 (2 marks) What are the 2 algebraic methods of solving simultaneous equations called?

_____

_____

2 How do you find the solution to simultaneous equations using their graphs?

_____

_____

_____

3 How can you check the solution of simultaneous equations?

_____

_____

4 (4 marks) Solve these simultaneous equations using the elimination method.

$$2x - 3y = 9$$
$$5x + y = 14$$

_____

_____

_____

_____

_____

_____

_____

_____

_____

_____

_____

(10) SIMULTANEOUS EQUATIONS 2

ELIMINATION AND SUBSTITUTION ARE THE 2 ALGEBRAIC METHODS OF SOLVING SIMULTANEOUS EQUATIONS.

Part A	/ 8 marks
Part B	/ 8 marks
Part C	/ 8 marks
Part D	/ 8 marks
Total	/ 32 marks

PART A: *MENTAL MATHS*

🚫 Calculators not allowed

1 Round $103°45'40"$ to the nearest minute.

_____

2 Simplify $(-2x^2y^0)^3$.

_____

3 Expand $(a + 5)(2a + 3)$. _____

_____

_____

4 Find the surface area of this rectangular prism.

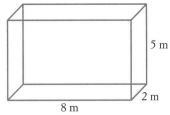

5 m

2 m

8 m

_____

5 Calculate the time difference between 9:15 a.m and 5:20 p.m.

_____

6 Simplify each ratio.

a 5 mm : 2 m _____

b 3 h : 10 min _____

7 Factorise $4y^2 - 24y + 32$.

_____

PART B: *REVIEW*

Solve each pair of simultaneous equations using the elimination method.

1 $2a + b = 23$

$3a - 2b = 3$

_____

_____

_____

_____

_____

_____

_____

2 $3m - 2n = 19$

$4m + 7n = 6$

_____

_____

_____

_____

_____

_____

_____

_____

9780170454568

PART C: PRACTICE

> › The substitution method
> › Simutaneous equations problems

1 (4 marks) Solve this pair of simultaneous equations using the substitution method.

$y = 10 + 3x$

$y - -x + 6$

_____

_____

_____

_____

_____

_____

_____

2 (4 marks) It costs 4 adults and 3 children $500 for tickets to a show, while the cost of 3 adults and 2 children is $360. Use simultaneous equations to find the cost of each adult and child ticket.

_____

_____

_____

_____

_____

_____

PART D: NUMERACY AND LITERACY

1 When solving simultaneous equations, Melody made y the subject of one of the equations. Which method of solving simultaneous equations was she using?

_____

2 (3 marks) Put in order the following steps for solving a problem requiring simultaneous equations.

A Solve the problem by answering in words.

B Solve the equations.

C Write simultaneous equations for the problem.

D Identify the variables to be used.

E Read the problem carefully.

_____

3 (4 marks) Solve these simultaneous equations.

$$x = -2 + 3y$$
$$y = x - 6$$

_____

_____

_____

_____

_____

(11) STARTUP ASSIGNMENT 11

> WE'LL BE COVERING QUADRATIC EQUATIONS AND THE PARABOLA IN THIS TOPIC.

PART A: BASIC SKILLS   / 15 marks

1 Calculate cos 45°, correct to 3 significant figures. _____

2 Evaluate $4^{\frac{1}{2}} \times 4^{-2}$. _____

3 Complete: 10 m³ = _____ cm³.

4 Complete: 55 m/s = _____ km/h.

5 Write a possible equation for this line.

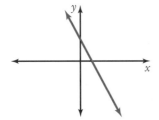

_____

6 Simplify $\dfrac{20x^3 y^2}{4x^3 y}$. _____

7 Find the surface area of a rectangular prism that has dimensions 4.6 cm, 2.1 cm and 5.0 cm.

_____

_____

8 Evaluate $\dfrac{2.46 \times 10^{12}}{4 \times 10^8}$. _____

9 Name the quadrilateral that has equal diagonals that bisect each other.

_____

10 If $\dfrac{5}{6}$ of a number is 90, what is the number?

_____

_____

11 A number from 1 to 20 is chosen at random. What is the probability that it is a multiple of 3?

_____

12 Find θ to the nearest degree if sin θ = 0.561.

_____

13 Divide $4550 in the ratio 2 : 1 : 4.

_____

14 What is the sum of the exterior angles of any polygon?

_____

15 Madeline is paid $465 plus a 5.5% commission on any jewellery she sells. Calculate her pay if she sells $3812 worth of jewellery.

_____

PART B: ALGEBRA AND GRAPHS   / 25 marks

16 For the line $y = 3x - 2$, find:

a its gradient _____

b its y-intercept. _____

17 If $x = 3$, find y if:

a $y = x^2 + 5$ _____

b $y = 2x^2 - x + 7$ _____

c $y = (x - 4)(x + 2)$ _____

d $y = 3x - x^2$ _____

18 Graph $y = 3x^2$ on a number plane.

24 Graph $y = -\dfrac{1}{2}x^2$ on a number plane.

25

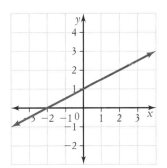

19 If $h = 3t^2 + 7t + 10$, find h if:

   a $t = 2$ _____

   b $t = -2$ _____

   c $t = 5$ _____

20 Solve:

   a $x^2 - 5x - 14 = 0$ _____

      _____

   b $y^2 - 4y - 5 = 0$ _____

      _____

21 Expand $(x - 3)^2$.

_____

22 Factorise $3x^2 + 11x - 20$.

_____

_____

23 If $x = -2$, find y if:

   a $y = x^2 - 4$ _____

   b $y = -x^2 + 4x - 3$ _____

   c $y = 4x^2 - 6x$ _____

For this line, find:

   a its y-intercept _____

   b its x-intercept _____

   c its gradient _____

   d its equation. _____

26 If $a = -5$ and $b = -2$, evaluate $-\dfrac{b}{2a}$.

_____

27 Evaluate $\sqrt{b^2 - 4ac}$, correct to 2 decimal places, if:

   a $a = 2$, $b = 4$ and $c = 1$ _____

   b $a = -1$, $b = -3$ and $c = 4$ _____

PART C: CHALLENGE   Bonus   / 3 marks

What is the value of $\sqrt{2 + \sqrt{2 + \sqrt{2 + \sqrt{2 + \ldots}}}}$?

There is a definite answer. Can you prove it?

Hint: Let $x = \sqrt{2 + \sqrt{2 + \sqrt{2 + \sqrt{2 + \ldots}}}}$ and square both sides. _____

⑪ QUADRATIC EQUATIONS PUZZLE

THE SOLUTIONS TO THESE EQUATIONS CAN BE FOUND NEXT PAGE. MATCH THEM TO SOLVE THE RIDDLE.

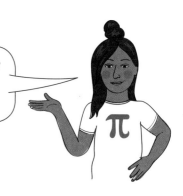

PUZZLE SHEET

PS

| 12 | 29 | 23 | 26 | 25 | 30 | 2 | | 24 | 9 | 18 | 19 | | 2 | 23 | 11 | 22 | | 18 | 19 |

| 29 | 25 | 27 | 7 | | 2 | 28 | 13 | 12 | 9 | 18 | 30 | 2 | | 29 | 25 | 30 | 7 | 19 | | 23 | 30 | 10 |

| 6 | 15 | 28 | 8 | 7 | 19 | | 21 | 30 | | 13 | | 30 | 15 | 11 | 20 | 7 | 28 | | 12 | 29 | 23 | 30 | 7 | .

| 26 | 5 | 15 | , | 8 | 22 | | 2 | 21 | 24 | | 16 | 5 | | 29 | 21 | 5 | 27 | | 14 | 5 | 28 |

| 16 | 9 | 22 | | 25 | 30 | 24 | 22 | 28 | 6 | 7 | 12 | 16 | 19 | .

The numbers in the grid above match the question numbers. Solve each quadratic equation and match the solution with an answer from the 'Key' provided on the next page. Fill in the grid above with the letters that match the questions, to discover the answer to the riddle:

What did the maths coach say to his team of football players?

1 $(x - 3)(x - 2) = 0$

_____

2 $(2x - 3)(3x + 1) = 0$

_____

3 $x^2 - 16 = 0$

_____

4 $2x^2 - 18 = 0$

_____

5 $x^2 - 2 = 0$

_____

6 $x(2x + 5) = 0$

_____

7 $x^2 = 3x$

_____

8 $x^2 - 6x + 5 = 0$

_____

9 $2x^2 - 10x = 0$

_____

10 $3x^2 - 5x - 2 = 0$

_____

11 $6x^2 - 5x - 6 = 0$

_____

12 $4x^2 - 13x - 12 = 0$

_____

13 $x^2 + 7x - 10 = 0$

_____

14 $3x^2 - 11x + 6 = 0$

_____

15 $x(x + 2) + 3(x + 2) = 0$

_____

9780170454568

16 $\dfrac{x}{3} = \dfrac{12}{x}$

_____

17 $(x + 2)^2 = 5$

_____

18 $4x^2 + 2x - 6 = 0$

_____

19 $x^2 - 8x + 2 = 0$

_____

20 $\dfrac{16}{x - 3} = x + 3$

_____

21 $2x^2 + x - 5 = 0$

_____

22 $3x(x - 1) = 2(5 - 2x)$

_____

23 $(2x - 1)(3x + 1) = 4$

_____

24 $2x^2 + 12x + 10 = 0$

_____

25 $(3x - 1)^2 = 100$

_____

26 $\dfrac{x - 3}{7} = \dfrac{4}{x}$

_____

27 $6x - 8x^2 = 1$

_____

28 $3x^2 - 6x + 2 = 0$

_____

29 $x(2x - 3) = -1$

_____

30 $x = \dfrac{5x + 14}{x}$

_____

Key

A	$x = -\dfrac{5}{6}, 1$	**G**	$x = \dfrac{3}{2}, -\dfrac{1}{3}$	**N**	$x = 7, -2$	**T** $x = \pm 6$
A	$x = \dfrac{-7 \pm \sqrt{89}}{2}$	**H**	$x = 0, 5$	**O**	$x = \pm \sqrt{2}$	**U** $x = -3, -2$
B	$x = \pm 5$	**I**	$x = 1, -\dfrac{3}{2}$	**O**	$x = \dfrac{-1 \pm \sqrt{41}}{4}$	**V** $x = 1, 5$
C	$x = 0, -\dfrac{5}{2}$	**I**	$x = \dfrac{11}{3}, -3$	**P**	$x = 4, -\dfrac{3}{4}$	**W** $x = 2, 3$
D	$x = -\dfrac{1}{3}, 2$	**J**	$x = -2 \pm \sqrt{5}$	**Q**	$x = \pm 4$	**X** $x = \pm 3$
E	$x = -2, \dfrac{5}{3}$	**K**	$x = \dfrac{1}{4}, \dfrac{1}{2}$	**R**	$x = \dfrac{3 \pm \sqrt{3}}{3}$	**Y** $x = -4, 7$
E	$x = 0, 3$	**L**	$x = \dfrac{1}{2}, 1$	**S**	$x = 4 \pm \sqrt{14}$	
F	$x = \dfrac{2}{3}, 3$	**M**	$x = -\dfrac{2}{3}, \dfrac{3}{2}$	**T**	$x = -5, -1$	

THE ANSWERS ARE NEXT PAGE, BUT
THE LETTERS ARE SCRAMBLED.

Unscramble the word clues and write them in the crossword.

Across

2 TSUNAMILOUSE

4 SEORIFATC

5 TESOCONRDIA

9 STNNACOT

11 CACOEVN

14 MRTMYSYE

16 FLUROAM

17 BALARAPO

19 TIPCENTRE

21 STULOION

23 REECTNIOINST

Down

1 CIEFFOCEITN

3 AADIQCURT

6 CONIM

7 TORO

8 SERQUA

9 LETMECOP

10 XRVETE

12 ANIQUOTE

13 BYHPAOLER

14 NBSTIUSUTTOI

15 RUSD

18 SXAI

20 TAXCE

22 RCCLEI

(11) QUADRATIC EQUATIONS

THE QUADRATIC FORMULA IS THE LONGEST FORMULA YOU'LL EVER LEARN!

Name:

Due date:

Parent's signature:

Part A	/ 8 marks
Part B	/ 8 marks
Part C	/ 8 marks
Part D	/ 8 marks
Total	/ 32 marks

PART A: MENTAL MATHS

🚫 Calculators not allowed

1 An interval AB joins $A(-8, 5)$ and $B(4, 0)$ on the number plane. Find:

a its length

b its gradient

2 (4 marks) Factorise each expression.

a $x^2 + 3x - 88$

b $5n^3 - 125n$

3 For this set of data, find:

$$6 \quad 9 \quad 9 \quad 11 \quad 13 \quad 14 \quad 15 \quad 16 \quad 19$$

a the median

b the interquartile range

PART B: REVIEW

1 Solve each equation.

a $(x + 4)(x - 6) = 0$

b $x(2x - 9) = 0$

2 (6 marks) Solve each equation.

a $(2u - 7)(3u + 5) = 0$

b $4k^2 - 12k = 0$

9780170454568

c $7p^2 + 11p - 6 = 0$

_____

_____

_____

PART C: *PRACTICE*

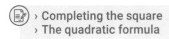 › Completing the square
› The quadratic formula

1 Solve each equation, writing the solution in surd form.

a $(x + 3)^2 = 5$

_____

_____

_____

b $\left(y - \dfrac{1}{2}\right)^2 = \dfrac{3}{4}$

_____

_____

_____

2 (4 marks) Solve each equation by completing the square.

a $a^2 + 2a - 5 = 0$

_____

_____

_____

b $2b^2 + 4b - 6 = 0$

_____

_____

_____

3 (2 marks) Solve $3x^2 + x - 7 = 0$ using the quadratic formula.

_____

_____

_____

_____

PART D: *NUMERACY AND LITERACY*

1 Find the numbers that 'complete the square' in this equation.

$$x^2 - 10x + \underline{\qquad} = (x - \underline{\qquad})^2$$

2 (6 marks) Solve each equation.

a $(2x - 5)^2 = 10$

_____

_____

_____

b $5x^2 + 3x - 2 = 0$

_____

_____

_____

c $3x^2 = 8x + 3$

_____

_____

_____

FOR QUESTION 4, THE SURFACE AREA OF A SPHERE IS $4\pi r^2$, BUT THE SURFACE AREA OF A HEMISPHERE IS NOT $2\pi r^2$.

Name:	
Due date:	
Parent's signature:	

Part A	/ 8 marks
Part B	/ 8 marks
Part C	/ 8 marks
Part D	/ 8 marks
Total	/ 32 marks

PART A: MENTAL MATHS

▦ Calculators not allowed

1 (2 marks) Factorise $6a^2 + 11a + 3$.

_____

_____

_____

2 Make x the subject of the formula $2y = \sqrt{\dfrac{x}{5}}$.

_____

_____

_____

3 Solve $-2m + 10 < -6$.

_____

_____

4 The formula for the surface area of a sphere is $SA = 4\pi r^2$. Find the surface area of this hemisphere in terms of π.

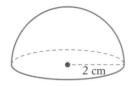

_____

5 (2 marks) Solve these simultaneous equations.

$$-3x + 4y = 20$$

$$6x - 5y = -19$$

_____

_____

_____

_____

6 Simplify $\sqrt{147} - \sqrt{75}$.

_____

_____

PART B: REVIEW

1 (3 marks) This right-angled triangle has an area of 600 m². Its base is 10 m longer than its height. Find its base and height.

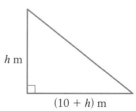

_____

_____

9780170454568

2 (2 marks) Solve $3x^2 - 3x - 6 = 0$.

_____

_____

_____

3 (3 marks) The product of 2 consecutive even numbers is 1848. Find the numbers.

_____

_____

_____

_____

PART C: PRACTICE

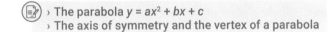
  › The parabola $y = ax^2 + bx + c$
  › The axis of symmetry and the vertex of a parabola

1 For this parabola, find:

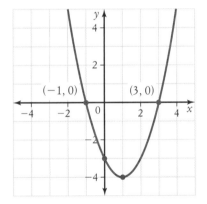

a the equation of the axis of symmetry

_____

b the coordinates of the vertex _____

2 (6 marks) Graph $y = -2x^2 + 6x - 4$, showing its vertex, x- and y- intercepts.

PART D: NUMERACY AND LITERACY

1 (2 marks) Is the graph of $y = x^2 - 6x + 8$ concave up or concave down? Give reasons.

_____

_____

_____

2 Find the y-intercept of the graph of $y = x^2 - 6x + 8$.

_____

3 Find the equation of the axis of symmetry of the graph of $y = x^2 - 6x + 8$.

_____

_____

4 Find the coordinates of the vertex of the graph of $y = x^2 - 6x + 8$.

_____

_____

5 (2 marks) Find the x-intercepts of the graph of $y = x^2 - 6x + 8$.

_____

_____

_____

6 Graph $y = x^2 - 6x + 8$.

(11) GRAPHING PARABOLAS 2

SOME MORE COMPLEX
PARABOLAS TO GRAPH HERE.

Teacher's tickbox

Graph the ticked set of quadratic equations.

❏ $y = x^2, y = 2x^2, y = \frac{1}{2}x^2, y = -2x^2$ ❏ $y = x^2 + 4, y = 2x^2 - 2$

❏ $y = x^2 - 2x - 3, y = -x^2 + 5x - 6$ ❏ $y = \frac{1}{2}x^2 - 4x + 1, y = -x^2 - 4x + 5$

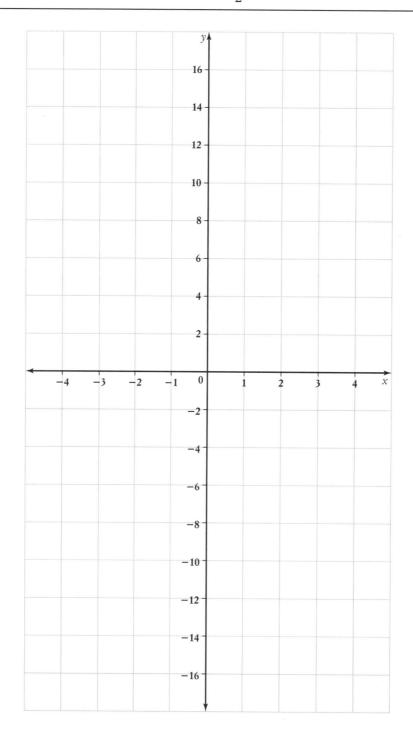

STARTUP ASSIGNMENT 12 ⑫

BEFORE WE START PROBABILITY, LET'S REVISE OUR CHANCE SKILLS.

PART A: BASIC SKILLS   / 15 marks

1 Complete: $1 \text{ m}^3 = $ _____ cm^3

2 Simplify $\dfrac{2a}{5} + \dfrac{a}{3}$. _____

3 Write one property of the diagonals of a rectangle. _____

4 Divide $112 in the ratio 5 : 2. _____

5 Calculate, correct to 2 decimal places, the volume of a cylinder with diameter 6.6 cm and height 11 cm. _____

6 Convert a salary of $47 800 to a weekly pay. (Answer to the nearest dollar.) _____

7 Find the value of y in the diagram below.

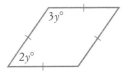

_____

8 Find the gradient of a line parallel to $y = 3x - 6$. _____

9 If $y = \dfrac{k}{x}$, find k if $y = 7$ when $x = 10$.

_____

10 Solve $3d - 4 = d + 10$. _____

11 A cap sells for $22.54 after an 8% discount. What was its original price?

_____

12 Calculate to the nearest $100 the value of a $29 000 car after 6 years if it depreciates at 16% per annum.

_____

13 Find the value of r.

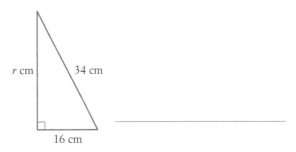

_____

14 A cube has a volume of 13.824 m³. Find its:

a length _____

b surface area. _____

PART B: PROBABILITY   / 25 marks

15 If the chance that it does not rain on Saturday is 42%, what is the chance that it does rain?

_____

16 What fraction of the numbers from 1 to 20 are:

a less than 8? _____

b square numbers? _____

c divisible by 4? _____

17 A ticket is drawn at random from a box containing 7 blue, 10 white, 4 red and 3 yellow tickets. Find the probability the ticket drawn is:

a red _____

b blue or white _____

c not red. _____

18 What is the probability that a person chosen at random has a birthday in a month beginning with M?

_____

19 Sort these events in order from most to least likely:

D rolling an odd number on a die

L buying your lunch at the canteen tomorrow

Y the next person visiting the class being in Year 9

P a letter arriving in the post for you today

I you watch YouTube tomorrow

S you are at school before 8:30 a.m. tomorrow

_____

20 Evaluate:

a $1 - 0.45$ _____

b $1 - \dfrac{5}{6}$ _____

21 What does it mean if an event has a probability of 0? _____

22 Write as a decimal the value of a '50-50 chance'. _____

23 What are the possible outcomes for the result of a soccer match between the Reds and the Blues?

_____

24 Henry, Irene, Jack, Kathy and Lisa wrote their names on separate cards. What is the probability that a card chosen at random has a boy's name?

_____

25 There are 400 tickets in a raffle. If Johanna buys 8 tickets, find the decimal probability she wins first prize. _____

26 If the probability of having 2 boys in a 3-child family is $\dfrac{3}{8}$, what is the probability of not having 2 boys? _____

27 A tossed coin came up heads 18 times and tails 22 times. What percentage of tosses showed tails? _____

28 What fraction of a deck of cards are:

a diamonds? _____

b aces? _____

c jacks, queens or kings? _____

d even numbers? _____

29 What percentage (to one decimal place) of the alphabet are vowels?

_____

PART C: CHALLENGE Bonus / 3 marks

Thomas, Patrick, Adrian, Alexis and Chris sit together in a row for a group photo.

How many possible seating arrangements are there?

_____

I'M ZINA. ALL THE ANSWERS TO THIS CROSSWORD ARE LISTED BELOW. YOU JUST HAVE TO WORK OUT WHERE THEY GO!

COMPLEMENTARY	COMPOUND	CONDITIONAL	DEPENDENT
DIAGRAM	DICE	DIE	EVENT
EXCLUSIVE	EXPECTED	EXPERIMENTAL	FREQUENCY
INDEPENDENT	LIST	MUTUALLY	OUTCOME
OVERLAPPING	PROBABILITY	RANDOM	RELATIVE
REPLACEMENT	SAMPLE	STEP	TABLE
THEORETICAL	TREE	TRIAL	TWO-WAY
VENN	WITHOUT		

12 TREE DIAGRAMS

TREE DIAGRAMS ARE OFTEN NEEDED FOR PROBABILITY PROBLEMS. MAKE SURE YOU PRACTISE DRAWING AND USING THEM.

1 **a** Explain what this tree diagram shows about two-child families.

First child	Second child	Outcomes
B	B	BB
B	G	BG
G	B	GB
G	G	GG

B = boy, G = girl

_____

_____

_____

_____

b Find the probability of a two-child family having:

i 2 girls _____

ii a boy followed by a girl _____

iii one boy and one girl born in any order.

_____

2 Piper sits a test in Science and in History.

a If she has an even chance of passing or failing each test, list all the possible outcomes on the tree diagram below.

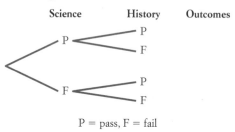

Science	History	Outcomes
P	P	
P	F	
F	P	
F	F	

P = pass, F = fail

b Find the probability that Piper:

i passes both tests _____

ii fails both tests. _____

3 Justin wants to work out the chance of rain over a long weekend (3 days).

a If rain (R) or no rain ($\bar{R}$) on each day are equally likely, complete the tree diagram below and list the outcomes.

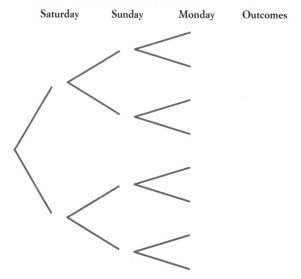

Saturday   Sunday   Monday   Outcomes

b Calculate the probability that, over the 3 days, it rains on exactly one of the days.

_____

4 Adrian and Pete each shoot for goal on a basketball court. Each player has an equal chance of scoring or missing the goal. Complete the tree diagram below to help you find the probability that at least one of them scores.

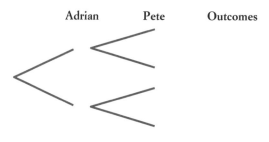

Adrian   Pete   Outcomes

_____

5 Anna rolls 2 dice and finds the difference between the 2 numbers rolled.

a Complete the table to show all the possible differences.

		1st die					
		1	**2**	**3**	**4**	**5**	**6**
2nd die	**1**	0					
	2						
	3						3
	4						
	5		2				
	6						

b Find the probability of rolling a difference:

i of 3 _____

ii of 0 _____

iii that is even _____

iv of more than 3. _____

6 A 4-sided die (numbered 1, 2, 3, and 4) is rolled and a coin is tossed.

a Use a tree diagram or table to list all possible outcomes.

b Find the probability of obtaining:

i a 2 and a head _____

ii an odd number and a tail _____

7 A bag contains blue, red and white socks. Three socks are chosen from the bag. Use a tree diagram to find the probability that at least 2 of the socks are white. _____

8 Which situation below can be illustrated by this tree diagram?

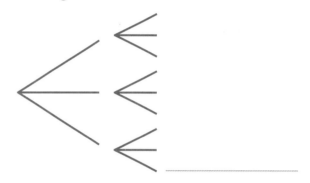

_____

A Tossing a coin 3 times

B Selecting 2 balls from a bag of red, blue and green balls

C 3 students passing or failing an exam

9 Georgina, Megan and Allyson sit in a row for a photo. What is the probability that:

a Megan sits in the middle? _____

b Allyson sits on either side? _____

c Georgina sits further left than Megan?

_____

12 TWO-WAY TABLES

A TWO-WAY TABLE IS A GREAT WAY OF DISPLAYING DATA ABOUT 2 TYPES OF CATEGORIES.

1 A sample of Year 7 students was surveyed on whether they owned a dog or cat.

	Have a dog	Do not have a dog
Have a cat	3	5
Do not have a cat	12	4

a How many students were surveyed?

_____

b How many students have a cat?

_____

c If a student is selected at random, what is the probability that he or she:

   i has a cat and a dog? _____

   ii has neither a cat nor a dog? _____

   iii does not have a cat? _____

2 This information was collected about the types of cars sold at a caryard.

	Automatic	Manual
Sedan	32	18
Hatchback	11	9

a How many cars were there? _____

b How many cars were sedans? _____

c If a car is selected at random, what is the probability it is:

   i an automatic? _____

   ii a manual hatchback? _____

d If a sedan is chosen at random, what is the probability it is a manual? _____

3 People in a shopping centre were surveyed on their favourite leisure activity.

	Dance	Sport	TV
Female	3	12	7
Male	15	8	5

a How many people were surveyed?

_____

b Which activity was the most popular?

_____

c What is the probability that a person chosen at random from the survey:

   i is female? _____

   ii said sport was their favourite activity?

_____

   iii is a male whose favourite activity is watching TV? _____

   iv is a female whose favourite activity is dance or sport? _____

4 A hospital kept statistics about the birthweight of babies born in the last month.

	Less than 5 kg	5 kg or more
Female	25	7
Male	19	9

a How many girls were born? _____

b How many babies weighing 5 kg or more were born? _____

c What is the probability that a baby chosen at random from this hospital:

i is male? _____

ii weighs less than 5 kg? _____

iii is female and weighs 5 kg or more?

_____

5 Year 9 students were surveyed on how they normally travel to school.

	Male	Female
Walk	0	0
Car	13	18
Bus	48	37
Cycle	7	2
Skateboard	5	0

a How many students were surveyed?

_____

b Were there more males or more females surveyed?

_____

c What is the probability that a Year 9 student selected at random:

i cycles to school? _____

ii is female and catches a bus to school?

_____

iii uses a car or bus to get to school?

_____

d Why do you think no-one walks to school?

_____

(12) PROBABILITY 1

REMEMBER THAT A PROBABILITY VALUE CAN BE WRITTEN AS A FRACTION, DECIMAL OR PERCENTAGE.

HW HOMEWORK

Name:

Due date:

Parent's signature:

Part A	/ 8 marks
Part B	/ 8 marks
Part C	/ 8 marks
Part D	/ 8 marks
Total	/ 32 marks

PART A: MENTAL MATHS

🚫 Calculators are not allowed

1 (4 marks) Find x and y, giving reasons.

$75°$ $x°$ $y°$

_____

_____

_____

_____

2 (2 marks) Solve $\dfrac{5x-2}{7} = \dfrac{3x+2}{4}$.

_____

_____

_____

_____

3 (2 marks) Find the centre and radius of the circle with equation $x^2 + y^2 = 16$.

_____

_____

PART B: REVIEW

1 The probability of rain tomorrow is 0.75. What is the probability of no rain tomorrow?

_____

2 A bag contains 2 red crayons, 3 yellow crayons and 5 blue crayons. A crayon is drawn at random from the bag and the colour recorded.

a How many colours are in the sample space?

_____

b Are the colours equally likely? Explain your answer.

_____

_____

_____

_____

3 (3 marks) This two-way table shows the survey results on students' favourite pets. Complete the table.

	Dog	Cat	Others	Total
Girls	10	18		35
Boys		10		
Total	25		16	

4 James bought 5 tickets in a raffle in which 100 tickets were sold. What is the percentage probability that James wins 1st prize?

_____

5 What is the expected frequency of tails if a coin is tossed 200 times?

_____

9780170454568

PART C: PRACTICE

› Relative frequency
› Venn diagrams
› Two-way tables

1 (2 marks) Lexie spun a spinner with 5 colours many times and recorded the results. Complete the table.

Colour	Frequency	Relative frequency
Yellow	20	$\frac{1}{5}$
Red	32	
Black	24	$\frac{6}{25}$
Blue	16	$\frac{4}{25}$
Green	8	

2 (2 marks) This Venn diagram shows the results of a survey on what types of pizza students like.

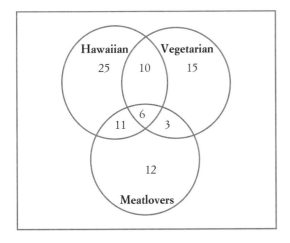

How many students like:

a both Vegetarian and Meatlovers?

_____

b Hawaiian or Vegetarian but not both?

_____

9780170454568

3 (4 marks) For 110 Year 10 students, 45 study Music, 56 study Art, and 24 do not study Music or Art. Show this information on a Venn diagram.

PART D: NUMERACY AND LITERACY

1 (2 marks) Complete:

$$P(E) = \frac{\text{number of favourable } _____}{\text{total number of } _____}$$

2 In the above formula, what does E stand for?

_____

3 What 2-word phrase describes the expected number of times an event will occur over repeated trials, such as the number of days it will rain in December?

_____

4 (4 marks) A group of students were asked about their favourite types of movies.

	Girls	Boys
Comedy	20	22
Horror	10	11
Action	14	34
Romance	30	9

a How many students were surveyed?

_____

b What is the probability of randomly selecting a student who:

i prefers comedy?

_____

ii is a boy who likes horror movies?

_____

iii is a girl who doesn't like action movies?

_____

(12) PROBABILITY 2

THIS PROBABILITY TOPIC IS GETTING QUITE HARD, SO YOU'LL NEED TO PRACTISE AND SHARPEN YOUR SKILLS.

Part A	/ 8 marks
Part B	/ 8 marks
Part C	/ 8 marks
Part D	/ 8 marks
Total	/ 32 marks

PART A: MENTAL MATHS

🚫 Calculators are not allowed

1 Convert 63.715° to degrees and minutes, to the nearest minute.

_____

2 Simplify $4a^{20}b^{35} \div 12a^4b^5$.

_____

3 Find:

a the scale factor for these similar figures

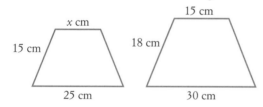

_____

b the value of x.

_____

_____

4 Find the shaded area in terms of π.

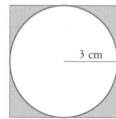

_____

_____

5 (2 marks) Solve $2(3y + 1) + 5 = 4(4y + 6)$.

_____

_____

_____

6 Write 25 365 000 in scientific notation.

_____

PART B: REVIEW

1 From a group of 945 people, what is the expected frequency of randomly selecting a person who was born on a Tuesday?

2 (5 marks) This Venn diagram shows whether 50 people liked cake or lollies.

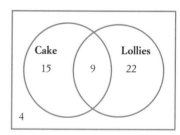

a How many people liked lollies? _____

b How many people did not like cake or lollies?

_____

c Complete this two-way table.

	Cake	Not cake
Lollies		
Not lollies		
Total		

3 (2 marks) 2 coins are tossed together 40 times and the number of heads counted each time.

Number of heads	Frequency
0	8
1	23
2	9
Total	40

What is the relative frequency of tossing:

a 2 heads? _____

b 2 tails? _____

PART C: *PRACTICE*

› Tree diagrams
› Dependent and independent events
› Conditional probability

1 (4 marks) 3 coins are tossed.

a Draw a tree diagram to show all possible arrangements of heads (H) and tails (T).

b How many outcomes are there in the sample space? _____

c Find the probability of tossing at least one head. _____

2 (3 marks) A bag contains 6 blue socks and 4 green socks. 2 socks are chosen randomly without replacement. Find the probability of choosing:

a a blue sock first

_____

b a green sock second after a green sock was chosen first

_____

3 2 dice are rolled. What is the probability of rolling a 3 on the second die, given that a 3 was rolled on the first die?

_____

PART D: *NUMERACY AND LITERACY*

1 What type of diagram has circles (usually overlapping) for grouping items into categories?

_____

2 How many different outcomes are possible when 2 dice are rolled? _____

3 (6 marks) 2 buttons are drawn without replacement from a bag containing 3 yellow and 2 blue buttons. Draw a tree diagram for the 20 possible outcomes and use it to find the probability that:

a both buttons are yellow _____

b the second button is blue _____

c the second button is yellow, given that the first button is yellow

_____

d both buttons are blue, given that both buttons are the same colour

_____

(13) STARTUP ASSIGNMENT 13

THIS ASSIGNMENT WILL HELP US TACKLE THE GEOMETRY TOPIC.

PART A: BASIC SKILLS   / 15 marks

1 Calculate, correct to 2 decimal places:

   a $5.7 \cos 42° 31'$ _____

   b the value of a $27 000 car after 2 years if it depreciates at 11% p.a. _____

   c the volume of a cylinder with radius 6 m and height 6 m _____

2 Describe the **mode** of a list of scores.

   _____

3 Find the scale on a map if 4 cm represents 200 m. _____

4 Find the average of b, $b + 10$ and $b - 1$.

   _____

5 What is the bearing of P from O?

   _____

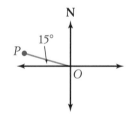

6 Solve $3(x + 7) = 5x + 14$. _____

7 Find the value of x in the diagram below.

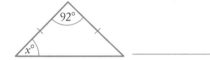

_____

8 Maddison walks at 6 km/h for $1\frac{3}{4}$ hours. How far does she walk? _____

9 Find the surface area of a cube with side length 4.5 cm. _____

10 What size angle does the minute hand of a clock turn in 4 minutes? _____

11 For $X(-2, 3)$ and $Y(-10, -12)$ on the number plane, find the length of XY. _____

12 Calculate Yasmin's pay if she earns $12.35 per hour for 8 hours plus $2\frac{1}{2}$ hours at time-and-a-half. _____

13 What is the gradient of $4x - 8y + 24 = 0$?

   _____

PART B: GEOMETRY   / 25 marks

14 Measure the size of $\angle JOK$ below.

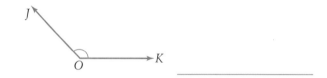

_____

15 Draw a rhombus and mark its axes of symmetry.

16 Which angle is vertically opposite $\angle CEB$ below?

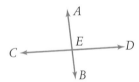

_____

17 What is the angle sum of a quadrilateral?

   _____

18 What type of triangle has all angles 60°?

   _____

9780170454568

19 A map's scale indicates that 1 cm represents 50 km.

a What distance does 5.5 cm on the map represent? _____

b 2 towns are 325 km apart. What is their scaled distance on the map?

_____

c Write the scale as a simplified ratio.

1 : _____

20 Draw an obtuse-angled triangle.

21 Find the value of each variable.

a

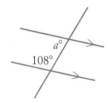

$a =$ _____

b

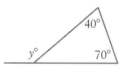

$y =$ _____

c

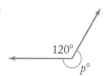

$p =$ _____

d

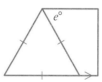

$e =$ _____

e

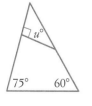

$u =$ _____

f

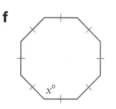

$x =$ _____

22 How many sides has an octagon? _____

23 True or false? The diagonals of a parallelogram:

a are equal _____

b bisect each other _____

c cross at right angles _____

24 If an isosceles triangle has a 130° angle, what are the sizes of the other 2 angles?

_____

25 a Which 2 triangles are congruent?

_____

b Which congruence test can be used for these triangles? _____

26 a If △ABC ||| △JLK, which angle matches ∠A?

 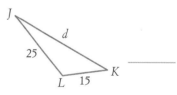

_____

b Find the value of d. _____

PART C: CHALLENGE   Bonus   / 3 marks

Place 5 crosses on this grid so that no 2 crosses are in the same row, column or diagonal.

(13) PROVING PROPERTIES OF QUADRILATERALS

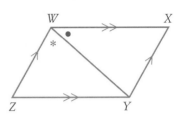

WE'RE USING CONGRUENT TRIANGLES TO PROVE PROPERTIES OF SPECIAL QUADRILATERALS.

1 The kite

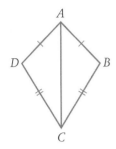

a If *ABCD* is a kite, which congruence test proves that △*ADC* ≡ △*ABC*?

_____

b Mark all pairs of matching angles on kite *ABCD*.

c Complete: One of the diagonals of a kite is an _____ of symmetry which _____ two angles of the kite.

d Draw diagonal *DB* on the kite to cross *AC* at *E*.

e Why is it true that △*ADE* = △*ABE*?

_____

f Which congruence test proves that △*ADE* ≡ △*ABE*? _____

g Which angle is equal to ∠*AED*?

_____

h What is the size of ∠*AED*?

_____

i Complete: The diagonals of a kite cross at

_____.

2 The parallelogram

a In this parallelogram, which congruence test proves that △*ZWY* ≡ △*XYW*?

_____

b Which angle is equal to ∠*WZY*?

_____

c Mark all pairs of matching angles on *WXYZ*.

d Complete: In a parallelogram, opposite angles are _____ and opposite sides are _____.

e Draw diagonal *ZX* to cross *WY* at *O*.

f Why is it true that ∠*WOZ* = ∠*XOY*?

_____

g Which congruence test proves that △*WOZ* ≡ △*XOY*? _____

h Mark all pairs of matching sides on *WXYZ*.

i Complete: The diagonals of a parallelogram _____ each other.

9780170454568

3 The rhombus

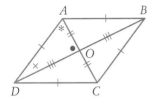

a In this rhombus, which test proves that
△AOD ≡ △AOB ≡ △COB ≡ △COD?

_____

b Mark all pairs of matching angles on *ABCD*.

c What are the sizes of the 4 angles at the centre
of the rhombus? _____

d Complete: In a rhombus, the diagonals bisect
at _____ and also _____ the
4 angles of the rhombus.

4 The rectangle

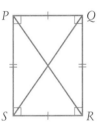

a If *PQRS* is a rectangle, which congruence test
proves that △*PSR* ≡ △*QRS*?

_____

b Which side of △*QRS* is equal to *PR*?

_____

c Complete: The diagonals of a rectangle
are _____ .

5 The square

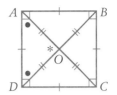

a A square is a special type of rectangle and
parallelogram. What does this mean about its
diagonals?

_____

b For square *ABCD*, which test proves that
△*AOD* ≡ △*AOB* ≡ △*BOC* ≡ △*DOC*?

_____

c Why is it true that ∠*OAD* = ∠*ODA*?

_____

d Mark all pairs of matching angles in square
ABCD.

e What is the size of ∠*OAD* and its matching
angles? _____

f What are the sizes of the 4 angles at the centre
of the square? _____

g Complete: The diagonals of a square are
_____ and bisect each other at
_____ _____. The diagonals also
_____ the right angles of the square.

(13) GEOMETRY CROSSWORD

THIS MATHEMATICAL SYMBOL, ≡, MEANS 'IS IDENTICAL OR CONGRUENT TO'.

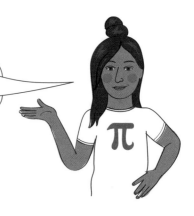

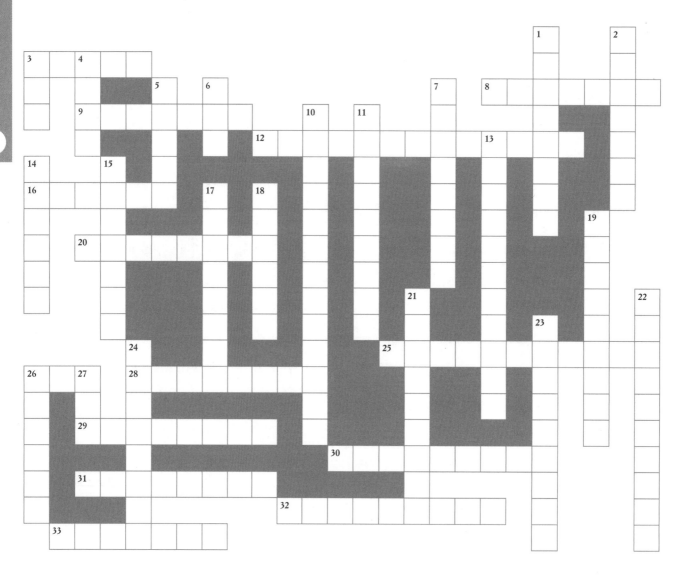

Clues across

3 Similar figures have sides whose lengths are in the same r_____

8 9-sided polygon

9 Same shape, different size

12 The name for a 4-sided polygon

16 To cut in half

20 In congruent figures, _____ sides are equal

25 The process of making a shape bigger

26 Side-side-side (abbreviation)

28 'Outside' angle of a triangle

29 A kite has one axis of _____

30 12-sided shape

31 The opposite of 28 across

32 The opposite of 25 across

33 Any 2D shape with straight sides

Clues down

1 5-sided military headquarters

2 A 'pushed-over' square

3 Right angle-hypotenuse-side (abbreviation)

4 There are 4 of these for proving congruent triangles

5 90° angle

6 Angle-angle-side (abbreviation)

7 Interval joining one vertex to an opposite vertex

10 Quadrilateral whose opposite angles are equal

11 Quadrilateral with one pair of parallel sides

13 Each angle in this triangle is 60°

14 Angle between 90° and 180°

15 10-sided polygon

17 An exterior angle of this equals the sum of the 2 interior opposite angles

18 The angle sum of a triangle is one hundred and _____ degrees

19 Triangle with 2 equal angles

21 Geometrical word for identical

22 This type of symmetry requires spinning a figure

23 Quadrilateral of right angles

24 Branch of mathematics that angles and shapes belong to

26 Diagonals bisect the angles of this shape at 45°

27 One of the congruent triangle tests (abbreviation)

(13) SIMILAR FIGURES

WHAT ARE THE 4 TESTS FOR SIMILAR TRIANGLES?

Name:	
Due date:	
Parent's signature:	

Part A	/ 8 marks
Part B	/ 8 marks
Part C	/ 8 marks
Part D	/ 8 marks
Total	/ 32 marks

PART A: MENTAL MATHS

🖩 Calculators not allowed

1 Simplify each ratio.

a 4 hours : 1 day

_____

b $2.80 : 20c

_____

2 Evaluate $2\frac{4}{5} \div 1\frac{1}{4}$. _____

_____

3 Simplify $-24a^2b^4 \div 4ab$.

_____

4 Find x and y.

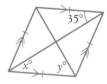

_____

5 For this data set, find:

37 36 40 41 38 38 42 36 39

a the mode

_____

b the median

_____

PART B: REVIEW

1 (6 marks) In $\triangle ABC$, $AB = AC$ and $AD \perp BC$.

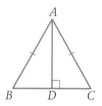

a Which congruence test can be used to prove that $\triangle ABD \equiv \triangle ACD$? _____

b Which angle matches with $\angle BAD$? _____

c Prove that $\triangle ABD \equiv \triangle ACD$ and hence show that AD bisects BC.

_____

_____

_____

_____

_____

_____

2 a Which similarity test proves that these triangles are similar? _____

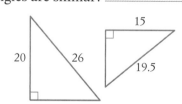

b What is the scale factor if the left triangle is the original? _____

PART C: PRACTICE

1 (6 marks) Find the value of each variable for each pair of similar figures.

a

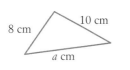
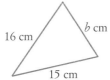

8 cm   10 cm   16 cm   b cm   15 cm

_____

_____

_____

b

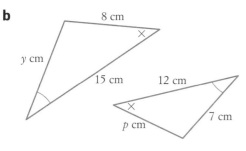

8 cm   y cm   15 cm   12 cm   p cm   7 cm

_____

_____

_____

c

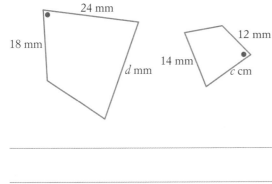

24 mm   18 mm   d mm   14 mm   12 mm   c cm

_____

_____

_____

2 Complete: The similarity test 'SSS' means that the 3 sides of one _____ are _____ to the 3 sides of the other _____.

PART D: NUMERACY AND LITERACY

1 Complete:

a For an enlargement, the scale factor is greater than _____.

b For similar figures, matching sides are in the same _____ and matching angles are _____.

2 a (5 marks) In the diagram below, why is $\angle E = \angle BCA$?

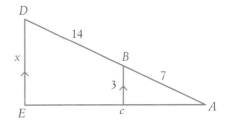

D   14   x   B   7   3   E   c   A

b There are 2 similar triangles in this figure. Complete: $\triangle ABC \; ||| \; \triangle$ _____.

c Which similarity test proves that these triangles are similar? _____

d Find the value of x.

_____

_____

THIS GEOMETRICAL SYMBOL, |||, MEANS 'IS SIMILAR TO'.

Name:

Due date:

Parent's signature:

Part A	/ 8 marks
Part B	/ 8 marks
Part C	/ 8 marks
Part D	/ 8 marks
Total	/ 32 marks

HW HOMEWORK

PART A: MENTAL MATHS

🚫 Calculators not allowed

1 Find 15% of $90.

_____

2 Simplify 46 : 58 : 28.

_____

3 A phone call costs $2.70 for 30 minutes. Express the cost as a rate in cents per minute.

_____

4 Factorise each expression.

a $x^2 - 2x - 24$

_____

b $-6a^2b^3 - 50ab$

_____

5 Simplify $\left(\dfrac{6y}{7x}\right)^{-2}$.

_____

6 Find the value of a and b, giving reasons.

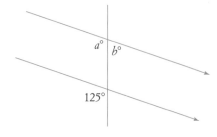

_____

_____

PART B: REVIEW

1 (4 marks) Test whether each pair of figures are similar, giving reasons.

a

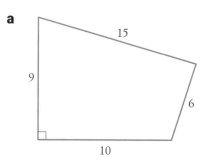

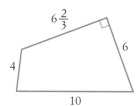

_____

_____

_____

_____

b

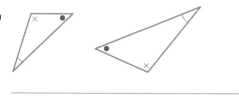

_____

_____

_____

9780170454568

2 Find the value of the variable in each pair of similar figures.

a

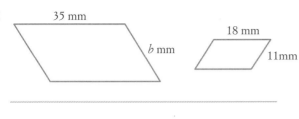

35 mm

b mm

18 mm

11mm

_____

_____

_____

b

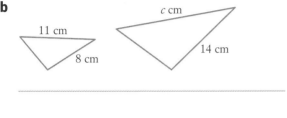

c cm

11 cm

8 cm

14 cm

_____

_____

_____

3 Identify the pair of similar figures below and the scale factor between them.

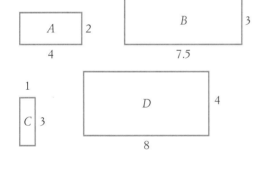

A 2

4

B 3

7.5

1

C 3

D 4

8

_____

_____

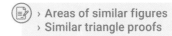
› Areas of similar figures
› Similar triangle proofs

1 Find the ratio of the areas of each pair of similar figures.

a

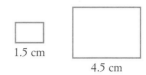

1.5 cm

4.5 cm

_____

_____

b

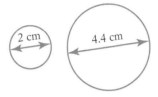

2 cm

4.4 cm

_____

_____

2 (2 marks) Two similar trapeziums have their areas in the ratio 49 : 36. Find the ratio of the lengths of their matching sides.

_____

3 (4 marks) Prove that each pair of triangles are similar.

a

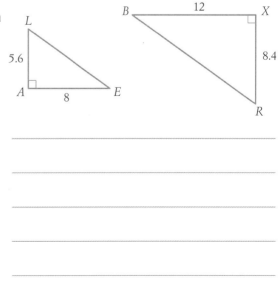

L

5.6

A 8 *E*

B 12 *X*

8.4

R

_____

_____

_____

_____

_____

HOMEWORK

HW

b

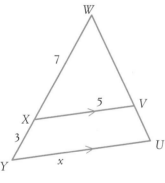

13.75

16

9.6

11

12.8

12

K

F

G

E

P

W

_____

_____

_____

_____

_____

PART D: NUMERACY AND LITERACY

1 (4 marks) Find the value of x after proving that 2 triangles are similar.

W

7

5

V

X

3

U

x

Y

_____

_____

_____

_____

_____

_____

_____

2 (4 marks)

a Prove that $\triangle FGE \;|||\; \triangle EFH$.

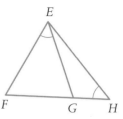

E

F

G

H

_____

_____

_____

_____

_____

_____

b If $FG = 16$ cm and $EF = 20$ cm, find the length of FH.

_____

_____

_____

9780170454568

1 $KL = LM$ and $PM = MN$. Find the size of $\angle N$.

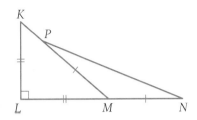

a $\angle KML = \dfrac{180° - 90°}{\square} = \underline{\hspace{1cm}}°$ (angle sum

of isosceles $\triangle$ \underline{\hspace{0.8cm}})

b $\therefore \angle \underline{\hspace{1cm}} = \underline{\hspace{0.8cm}}° - \underline{\hspace{0.8cm}}° = 135°$

(angles in a straight line)

c $\therefore \angle N = \dfrac{180° - \square°}{2} = \underline{\hspace{1cm}}° (\underline{\hspace{1.5cm}})$

2 $AC \parallel DG$. For $\triangle BFE$, prove that the exterior angle $\angle BFG$ is equal to the sum of the two interior opposite angles $\angle BEF$ and $\angle EBF$.

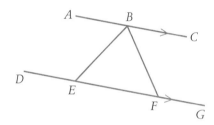

a Let $x = \angle ABE = \angle BEF$ (\underline{\hspace{1cm}} angles,

$AC \parallel \underline{\hspace{1cm}}$)

b Let $y = \angle EBF$

$\therefore \angle ABF = \underline{\hspace{1cm}} + \underline{\hspace{1cm}}$

c $\angle BFG = \angle \underline{\hspace{0.8cm}} (\underline{\hspace{0.5cm}}$ angles, $\underline{\hspace{0.5cm}} \parallel \underline{\hspace{0.5cm}})$

d $\therefore \angle \underline{\hspace{0.8cm}} = x + y$

e $\therefore \angle BFG = \angle \underline{\hspace{1cm}} + \angle \underline{\hspace{1cm}}$

3 A and B are the centres of 2 circles that intersect at C and D. Use congruent triangles to prove that BA bisects $\angle DAC$.

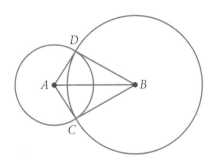

a In $\triangle DAB$ and $\triangle \underline{\hspace{1.5cm}}$

b $DA = \underline{\hspace{1.5cm}}$ (equal radii)

c $\underline{\hspace{1.5cm}} = \underline{\hspace{1.5cm}}$ (equal radii)

d AB is $\underline{\hspace{1.5cm}}$

e $\therefore \triangle \underline{\hspace{1.2cm}} \equiv \triangle \underline{\hspace{1.2cm}} (\underline{\hspace{1.2cm}})$

f $\therefore \angle DAB = \angle \underline{\hspace{1.5cm}}$ (matching angles in

$\underline{\hspace{1.5cm}}$ triangles)

g $\therefore \underline{\hspace{4cm}}$

4 $\angle PWT = \angle WNT = 90°$, $PN = 4$ cm, $NT = 9$ cm. Prove that $\triangle PWN$ is similar to $\triangle WTN$ and find the length of WN.

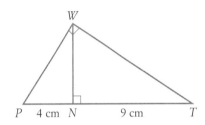

a Let $\angle P = \theta$

$\therefore \angle PWN = 180° - (90° - \theta) = \underline{\hspace{1.5cm}}$

(angle $\underline{\hspace{1.5cm}}$ of $\triangle \underline{\hspace{1cm}}$)

b Also $\angle TWN = 90° - \underline{\hspace{1.5cm}} = \underline{\hspace{1.5cm}}$

(angles in right angle $\angle PWT$)

c $\therefore \angle \underline{\hspace{1.5cm}} = \angle P = \theta$

d In $\triangle \underline{\hspace{1.5cm}}$ and $\triangle \underline{\hspace{1.5cm}}$

e $\angle WNP = \angle \underline{\hspace{1.5cm}} = \underline{\hspace{1.2cm}}°$ (given)

f $\angle P = \angle \underline{\hspace{1.5cm}} = \theta$ (proven above)

g $\therefore \triangle PWN \parallel\parallel\parallel \triangle \underline{\hspace{1.5cm}} (\underline{\hspace{1.5cm}})$

h $\therefore \dfrac{WN}{9} = \dfrac{\square}{\square} (\underline{\hspace{1.5cm}}$ sides in similar

triangles)

i $WN^2 = \underline{\hspace{1.5cm}}$

j $WN = \underline{\hspace{1.5cm}}$ cm

(14) STARTUP ASSIGNMENT 14

NOW WE'RE STARTING TO PREPARE FOR YEAR 11 HIGHER MATHS.

PART A: BASIC SKILLS / 15 marks

1 Find the standard deviation of these scores, correct to one decimal place:

3, 8, 5, 12, 7, 2 _____

2 Draw a bearing of 300°.

3 Expand:

a $(3p + 8)(3p - 8)$

_____

b $(4 - \sqrt{5})^2$

_____

4 Complete: 6.2 m³ = _____ cm³

5 Write the quadratic formula for solving $ax^2 + bx + c = 0$.

_____

6 Evaluate: $\dfrac{2.5 \times 10^5}{\sqrt{15\,625}}$ _____

7 Write a formula for the surface area of a hemisphere with radius r. _____

8 Find d. _____

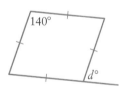

9 Which regular polygon has all angles 108°?

_____

10 Solve $\dfrac{15 - 5x}{3} \geq \dfrac{x + 1}{2}$.

_____

_____

_____

11 What is the probability that Deni rolls 2 6s on a pair of dice?

_____

12 An interval joins $(1, -4)$ and $(5, 0)$. Find:

a its midpoint _____

b its length as a simplified surd.

_____

_____

13 What type of graph has the equation $x^2 + y^2 = 1$? _____

PART B: GRAPHS / 25 marks

Answer the graphing questions on a separate page.

14 a Sketch on the same axes the graph of:

i $y = x^2$ **ii** $y = -x^2$

b What is the difference between these 2 graphs?

_____

_____

15 **a** Sketch on the same axes the graph of:

    i $y = 3x^2$ **ii** $y = 7x^2$ **iii** $y = \dfrac{1}{2}x^2$

b What is the difference between these 3 graphs?

    _____

    _____

    _____

16 Sketch the graph of:

a $y = \dfrac{1}{x}$ **b** $y = -\dfrac{1}{x}$

c $y = -x^2 + 1$ **d** $y = -x^2 - 4$

e $y = 3^x$ **f** $y = 3^{-x}$

17 Make x the subject of:

a $y = 3x + 4$

    _____

b $y = 5x^2$

    _____

c $y = -\dfrac{2}{x}$

    _____

18 Simplify the expression $x^2 + 2x - 5$ if:

a $x = 1$ _____

b $x = 0$ _____

c $x = -2$ _____

d $x = a$ _____

e $x = 4a$ _____

f $x = -3a$ _____

19 Sketch the graph of:

a $y = x$ **b** $y = x^3$

c $y = x^3 - 3$

PART C: CHALLENGE   Bonus   / 3 marks

Solve $8^{3x-4} = 32^{x+8}$.

14 FUNCTION NOTATION

SUBSTITUTE FOR *x* THE NUMBER OR VARIABLE INSIDE THE f().

1 Complete this table.

	$f(x) = 3 - x$	$f(x) = x^2 + 6x$	$f(x) = \dfrac{1}{3x-1}$	$f(t) = t^2 - \dfrac{t}{4}$	$f(t) = 2t^2 - 3t + 5$
$f(1)$					
$f(-2)$					
$f(0)$					
$f(5)$					
$f\left(2\dfrac{1}{2}\right)$					
$f(c)$					
$f(-c)$					
$f(c + 2)$					
$f(3 - c)$					
$f(p^2)$					

2 If $p(x) = 10 - 3x$ and $g(x) = \sqrt{3x + 1}$, find:

a $p(-4)$

b $g(10)$

c $[p(2)]^2$

d $p(5) + g(5)$

3 For each of the following functions, find the value of x if $f(x) = 7$.

a $f(x) = 12 - 5x$

b $f(x) = x^2 - 2$

c $f(x) = \dfrac{7}{x - 1}$

d $f(x) = \dfrac{x - 12}{4 - x}$

> A TRANSLATION IS A SHIFT IN ONE OR MORE DIRECTIONS.

Graph $y = f(x)$. Then graph the other functions on the same number plane.

1 $f(x) = 2^x$

   a $y = f(x) - 3$

   b $y = f(x) + 4$

   c $y = f(x - 3)$

3 $f(x) = 3 - x$

   a $y = f(x) - 1$

   b $y = f(x + 3)$

   c $y = f(x - 1) + 3$

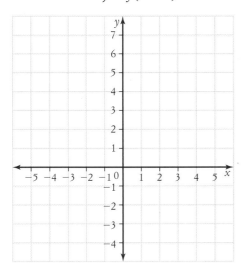

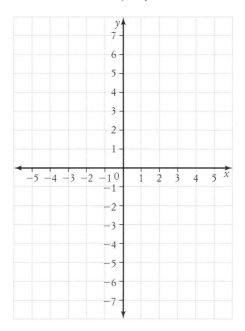

2 $f(x) = x^2$

   a $y = f(x) + 2$

   b $y = f(x + 2)$

   c $y = f(x - 4)$

4 $f(x) = \dfrac{2}{x}$

   a $y = f(x) + 3$

   b $y = f(x + 2)$

   c $y = f(x - 4)$

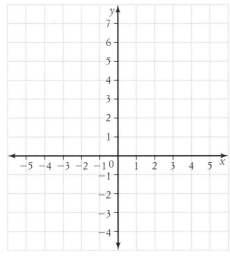

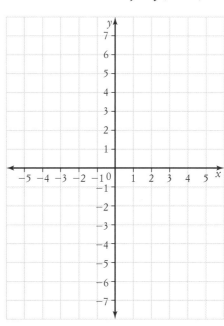

(14) POLYNOMIALS 1

Name:

Due date:

Parent's signature:

Part A	/ 8 marks
Part B	/ 8 marks
Part C	/ 8 marks
Part D	/ 8 marks
Total	/ 32 marks

THERE'S SOME HIGH-LEVEL ALGEBRA GOING ON HERE.

PART A: MENTAL MATHS

🚫 Calculators not allowed

1 (5 marks) Simplify each expression.

a $\dfrac{5x^2 - 14x - 3}{x^2 - 8x + 15}$

b $\dfrac{2x+4}{x^2+3x-4} \div \dfrac{x^2-4x-12}{x^2-2x-24}$

2 (2 marks) 2 similar triangles have heights in the ratio 2 : 5. If the smaller area is 100 cm², find the area of the larger triangle.

3 Simplify $\sqrt{300} + \sqrt{27}$.

PART B: REVIEW

1 If $x = 2$, evaluate each expression.

a $2x^3 - 3x^2 + x - 4$

b $3x^4 + 2x^3 - 4x$

2 (4 marks) Factorise each expression.

a $4x^3 - 64x$

b $-x^3 + 4x^2 + 60x$

3 (2 marks) Solve $3x^2 + 2x - 8 = 0$.

PART C: PRACTICE

› Dividing polynomials
› The remainder theorem

1 a What is the degree of the polynomial
$P(x) = 2x^3 - 2x + 5$?

_____

b What is the constant term?

_____

2 (4 marks) Divide $P(x) = 2x^3 - 2x + 5$ by
$A(x) = x - 4$, then write $P(x)$ in the form
$P(x) = A(x)Q(x) + R(x)$.

_____

_____

_____

_____

_____

_____

_____

_____

_____

3 Find the remainder when
$P(x) = 3x^3 - 4x^2 + 2x - 2$ is divided by:

a $(x - 1)$

_____

b $(x + 3)$

_____

PART D: NUMERACY AND LITERACY

1 Complete: A polynomial is an algebraic
expression involving powers of x that are

_____   _____.

2 a What is the leading coefficient of
$P(x) = x^6 + 4x^4 - 2x^3 + x - 1$?

_____

b Explain why $P(x)$ is a monic polynomial.

_____

_____

3 In the equation $P(x) = A(x)Q(x) + R(x)$, what
is the polynomial $R(x)$ called?

_____

4 (4 marks) Show that $(3x - 1)$ is a factor of
$P(x) = 27x^3 + 3x^2 - x - 1$, then express $P(x)$ as
a product of the 2 factors.

_____

_____

_____

_____

_____

_____

_____

HOMEWORK

HW

⑭ POLYNOMIALS 2

Name:

Due date:

Parent's signature:

Part A	/ 8 marks
Part B	/ 8 marks
Part C	/ 8 marks
Part D	/ 8 marks
Total	/ 32 marks

> IF YOU CAN DO THIS ASSIGNMENT, THEN YOU'LL BE WELL-PREPARED FOR YEAR 11 SPECIALIST MATHS.

PART A: MENTAL MATHS

🖩 Calculators not allowed

1 (2 marks) Find the gradient and y-intercept of the line with equation $-12x + 3y + 9 = 0$.

2 Write the linear equation $y = \dfrac{4}{7}x + 3$ in general form.

3 Find the time difference between 06:30 and 22:45.

4 (2 marks) Find in terms of π the surface area of a cylinder with a base radius of 5 cm and a perpendicular height of 8 cm.

5 (3 marks) Rationalise the denominator of each surd.

a $\dfrac{2}{3\sqrt{5}}$

b $\dfrac{4\sqrt{6}}{3\sqrt{3}}$

PART B: REVIEW

1 (2 marks) Write an example of a cubic polynomial that is monic and has a negative constant term.

2 (4 marks) Find the quotient and remainder when $x^5 - 2x^4 + x^3 - 2x^2 - 1$ is divided by $x + 1$.

3 Find the remainder when $5x^4 - 4x^2 + 3x - 2$ is divided by $x + 4$.

_____

4 Solve the equation $(x + 1)(x - 1)(x + 2) = 0$.

_____

_____

PART C: PRACTICE

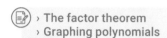

> The factor theorem
> Graphing polynomials

1 (2 marks) Show that $x + 1$ is factor of
$x^3 - 5x^2 + 2x + 8$.

_____

_____

_____

_____

2 (6 marks)

a Factorise $P(x) = -x^3 + 2x^2 + x - 2$ completely.

_____

_____

_____

_____

_____

b Hence find the x- and y-intercepts of the graph of $y = -x^3 + 2x^2 + x - 2$.

_____

_____

_____

_____

c Hence sketch the graph of
$y = -x^3 + 2x^2 + x - 2$.

PART D: NUMERACY AND LITERACY

1 (2 marks) Complete: If $P(x) = 0$ has a **double root** at $x = a$, then the graph of $y = P(x)$ at

$x = a$ _____ the x-axis with a

_____ gradient, like a parabola's vertex there.

2 (6 marks) Sketch the graph of each polynomial, showing x- and y-intercepts.

a $y = (x - 2)^2(x^2 - 1)$

b $y = -(x - 1)(x + 1)^3$

THIS ASSIGNMENT WILL PREPARE YOU FOR THE CIRCLE GEOMETRY TOPIC.

PART A: *BASIC SKILLS*   / 15 marks

1 Find the gradient of the line with equation

$2x - 4y - 5 = 0.$ _____

_____

_____

2 Find the exact value of tan 120°.

_____

3 Write the 4 tests for similar triangles.

_____

4 Solve $\dfrac{r + 4}{5} = \dfrac{2}{7}.$ _____

_____

_____

5 Calculate to 2 decimal places:

a the surface area of this cone

_____

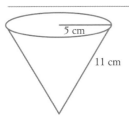

5 cm

11 cm

b the standard deviation of these scores:

16, 17, 24, 10, 16, 14, 13, 14, 19

_____

c the area of an equilateral triangle with 9.4 cm

sides.

_____

6 Ben buys 8 tickets in a raffle that has 500

tickets. Write as a percentage the probability

that he wins first prize. _____

7 Simplify $\left(\dfrac{2c^2}{d}\right)^{-3}.$ _____

8 Sketch the graph of $y = x^2 + 3.$

9 Calculate the simple interest on a loan of $4800

for 18 months at 12% p.a.

_____

10 Solve the simultaneous equations $x - 2y = 6$

and $3x + y = 11.$

_____

_____

_____

_____

_____

11 Find the equation of the line passing through

$(-3, 2)$ and $(6, 5).$

_____

_____

_____

_____

_____

12 If 2 similar solids have a scale ratio of 3 : 5,

what is the ratio of their surface areas?

_____

13 Calculate, to the nearest dollar, the value of a

$35 000 car after 2 years if it depreciates at 17% p.a.

_____

_____

_____

C
S
F

PART B: GEOMETRY

/ 25 marks

14 Draw and label on this circle:

a a segment

b a tangent.

15 Find the value of each variable.

a

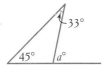

_____

_____

b

_____

_____

c

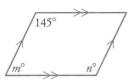

_____

_____

d

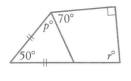

_____

_____

In questions **16**, **19** and **21**, O is the centre of the circle.

16 OC bisects the chord AB.

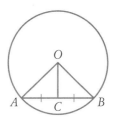

a Which test can be used to prove that

$\triangle OCA \equiv \triangle OCB$? _____

b Prove that $\triangle OCA \equiv \triangle OCB$.

_____

_____

_____

_____

_____

c Hence prove that $OC \perp AB$.

_____

_____

17 A rhombus has diagonals 16 cm and 30 cm.
Find the length of one side.

_____

_____

18 Calculate the area of this sector correct to
2 decimal places.

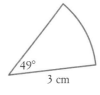

_____

_____

WORKSHEET

WS

19

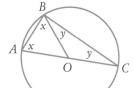

a Why does $\angle BAO = \angle ABO$?

_____

b Write an equation for the angle sum of $\triangle ABC$.

_____

_____

c Prove that $\angle ABC = 90°$.

_____

_____

20 Solve:

a $5x = 4 \times 7$

_____

_____

_____

b $3(d + 4) = 6 \times 16$

_____

_____

_____

c $y^2 = 6 \times 24$

_____

_____

_____

d $n(n + 5) = 3 \times 12$

_____

_____

_____

21

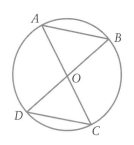

a Which test can be used to prove that
$\triangle AOB \equiv \triangle DOC$? _____

b Prove that $\triangle AOB \equiv \triangle DOC$.

_____

_____

_____

_____

_____

_____

c Hence prove that $AB = DC$.

_____

_____

d Why does $\angle A = \angle D = \angle B = \angle C$?

_____

_____

_____

e Hence prove that $AB \parallel DC$.

_____

_____

PART C: CHALLENGE   Bonus 3 marks

What fraction of the larger square's area is taken up by the smaller square?

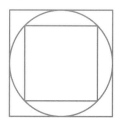

9780170454568

FIND THE VALUE OF EACH VARIABLE. WHEREVER IT APPEARS, O IS THE CENTRE OF THE CIRCLE.

1

2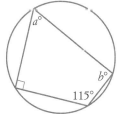

3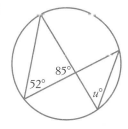

4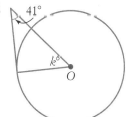

5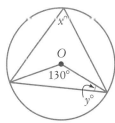

6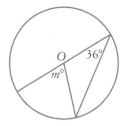

7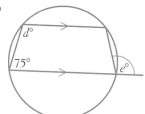

8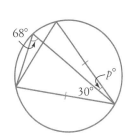

9 P is also a centre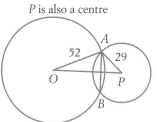

Find OP if AB = 40

10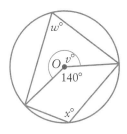

11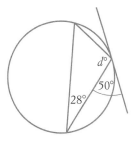

12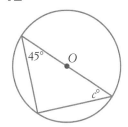

13

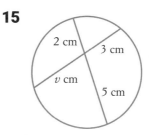

14

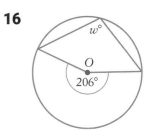

15

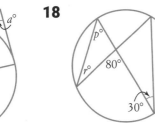

16

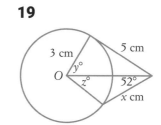

17

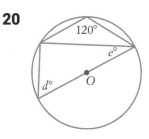

18

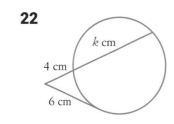

19

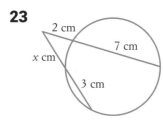

20

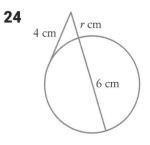

21

22

23

24

9780170454568

CONGRATULATIONS, YOU DID IT! REACHED THE END OF YEAR 10.

Name:

Due date:

Parent's signature:

Part A	/ 8 marks
Part B	/ 8 marks
Part C	/ 8 marks
Part D	/ 8 marks
Total	/ 32 marks

PART A: *MENTAL MATHS*

Calculators not allowed

1 Solve the equation $2x^2 - 4x - 5 = 0$.

_____

_____

_____

2 If the diagonals of a quadrilateral are equal in length and bisect each other, what special quadrilateral must it be?

_____

3 (2 marks) What is the centre and radius of the circle with equation $x^2 - 6x + y^2 + 5 = 0$?

_____

_____

_____

_____

_____

4 Find in terms of π the volume of this hemisphere.

_____

_____

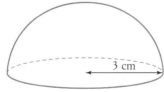

5 (2 marks) Simplify $\dfrac{1}{3\sqrt{5}} - \dfrac{2}{\sqrt{3}}$, giving the answer with a rational denominator.

_____

_____

_____

_____

6 Find the gradient of the line with equation $8x + 2y - 5 = 0$.

_____

_____

_____

PART B: *REVIEW*

Name each part of the circle labelled.

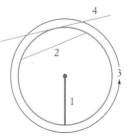

1 _____

2 _____

3 _____

4 _____

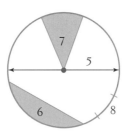

5 _____

6 _____

7 _____

8 _____

PART C: *PRACTICE*

(📝) › Chord properties of circles
› Angle properties of circles

1 (3 marks)

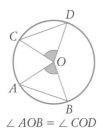

$\angle AOB = \angle COD$

a Prove that $\triangle AOB \equiv \triangle COD$.

_____

_____

_____

_____

_____

b Hence, prove that $AB = CD$.

_____

_____

2 (2 marks) Find x, giving reasons, if O is the centre of the circle.

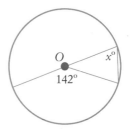

_____

_____

_____

3 (2 marks) Find m, giving reasons.

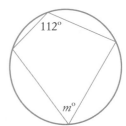

_____

_____

_____

4 O is the centre of this circle with radius 9 cm. Find, correct to 2 decimal places, the length of OA if the chord RS is 10 cm.

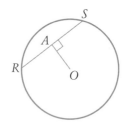

_____

_____

_____

_____

PART D: NUMERACY AND LITERACY

1 (2 marks) Complete: The exterior angle at a vertex of a cyclic quadrilateral is equal to _____ .

2 (6 marks) Find the value of each variable, giving reasons.

a

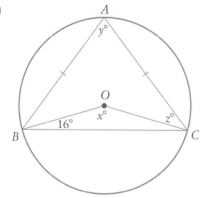

_____

_____

_____

_____

b

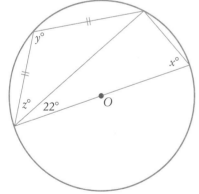

_____

_____

_____

_____

ANSWERS

Chapter 1

StartUp assignment 1 — PAGE 01

1 6.5

2 $3a + 7$

3 a 42.6 m **b** 77.4 m²

4 62 000

5 $625

6 2000

7 nP

8 $b = 30$

9 108°

10 $x = 11$

11 $\dfrac{1}{36}$

12 $(3, -2)$

13 x-axis

14 Circle, centre $(0, 0)$, radius 3 units

15 7

16 3.61

17 2

18 $\sqrt{25}, 0.1\dot{6}, 3.51$

19 $2x^2 + 5x$

20 $18a^4$

21 8 and 9

22 30

23 $3pr - 3p^2$

24 a 16, 4 **b** 25, 49

25 $h = \sqrt{13}$

26 $2x^2 + 7x - 4$

27 $\sqrt{40}, \sqrt{81}, \sqrt{98}, 10.1, 3.6^2$

28 a

29 a 25, 3 **b** 16, 5 or 4, 20

30 No

31 $4k^2 - 25$

32 Yes

33 a $x = \sqrt{3}$ **b** $\dfrac{\sqrt{3}}{2}$

34 $9x^2 - 6xy + y^2$

35 Yes

36 $\sqrt{58}$

Challenge: $x = \sqrt{2}$

Simplifying surds — PAGE 03

3.141 592 653 589 793 238 462 643

Rationalising the denominator — PAGE 04

1 $\dfrac{5\sqrt{2}}{2}$

2 $\dfrac{\sqrt{6}}{2}$

3 $\dfrac{\sqrt{15}}{3}$

4 $\dfrac{\sqrt{2}}{2}$

5 $\sqrt{6}$

6 $\dfrac{5\sqrt{30}}{12}$

7 $\dfrac{2\sqrt{5} + \sqrt{15}}{5}$

8 $\dfrac{\sqrt{30} - 3\sqrt{6}}{6}$

9 $2\sqrt{3} + 2$

10 $\dfrac{7\sqrt{5} - 5}{15}$

11 $2\sqrt{10} - \sqrt{6}$

12 $\sqrt{70} + 3$

13 $\dfrac{4\sqrt{2} - \sqrt{3}}{2}$

14 $\dfrac{7\sqrt{10} + 15\sqrt{2}}{30}$

15 $\dfrac{30 - 6\sqrt{2}}{23}$

16 $\sqrt{5} + 1$

17 $\dfrac{3\sqrt{2} - 2\sqrt{3}}{2}$

18 $\dfrac{5\sqrt{6} + \sqrt{30}}{5}$

19 $2\sqrt{6} + 3\sqrt{2}$

20 $\dfrac{13 + 5\sqrt{2}}{7}$

Surds crossword — PAGE 05

Across

5 difference

8 square

9 approximate

13 rationalise

15 root

16 real

Down

1 denominator

2 expression

3 binomial

4 undefined

6 irrational

7 quotient

8 simplify

10 product

11 expand

12 rational

14 surd

Chapter 2

StartUp assignment 2 — PAGE 10

1 33.909

2 a $64x^6$ **b** $\dfrac{8}{27}$

3 $10d + 6$

4 Volume of a cylinder

5 $x = \pm 4$

6 a $2 : 7$ **b** $6 : 5$

7 SAS

8 5.4 L/hour

9 $\sqrt{20}$ or $2\sqrt{5}$ units

10 65 cm²

11 No

12 $r = 69$

13 1.6×10^{11}

14 a 30 **b** 130

   c 3 **d** 16

   e 0.75 **f** 1.75

15 a $\dfrac{3}{25}$ **b** $\dfrac{1}{8}$

16 a 0.09 **b** 0.065

   c 0.015 **d** 0.008

17 23%

18 a $320 **b** $3.21

   c $1725 **19** $5382

20 $864 **21** $144.45

22 $600 **23** $86.50

24 a $2100.15 **b** $3254.86

25 $756 **26** $2349

Challenge: At the end of 12 years

Compound interest — PAGE 12

1 a 48 **b** 84

   c 42 **d** 15

2 a 0.04 **b** 0.065

   c 0.1325 **d** 0.0803

   e 0.075 **f** 0.0375

   g 0.006 **h** 0.001 48

3 a 0.011 25 **b** 0.015

   c 0.002 625 **d** 0.009 225

4 a 0.000 411 0 **b** 0.000 205 5

   c 0.000 591 8 **d** 0.000 458 1

5 a 0.045 **b** 0.0175

6 a $5832 **b** $6802.44

   c $7346.64

7 a $4908.93 **b** $19 539.18
  c $32 331.19 **d** $12 989.19
  e $22 721.51 **f** $52 294.04

8 a $1049.91 **b** $13 928.37
  c $6384.43 **d** $1084.20
9 a $2470.99 **b** $7166.79
  c $12 860.09

10

	Principal	Rate (% p.a.)	Time	Compounded	Final amount	Interest
a	$5500	7%	4 years	Yearly	$7209.38	$1709.38
b	$6372.75	6.4%	6 years	Half-yearly	$9300	$2927.25
c	$20 000	12.6%	3.5 years	Monthly	$31 013.83	$11 013.83
d	$21 397.28	9%	2 years	Monthly	$25 600	$4202.72
e	$8194.17	14.8%	$2\frac{3}{4}$ years	Quarterly	$12 220	$4025.83

Depreciation
PAGE 14

1 $28 343.52
2 a $13 384.40 **b** $20 615.60
3 $9686.44 **4** $198 766.99
5 a $277 992 **b** $216 833.76 **c** $61 158.24
6 a $8439.78 **b** 43.1%
7 a 90% **b** 81% **c** 59%
8 a $8162.10 **b** $6211.84 **c** 7 years
9 a $31 993.30 **b** $26 285.13 **c** 13 years
10 profit **11** 8 years
12 Disagree, the amount of depreciation changes each year because it is calculated from the previous year's value.

Interest and depreciation crossword
PAGE 16

Down

1 repayment **4** investment
6 gross **7** wage
10 principal **11** loading
12 fortnightly **16** rate
18 annual **20** term
22 compound **23** final
25 leave **26** salary

Across

2 simple **3** flat
5 overtime **8** tax
9 deposit **13** commission
14 deduction **15** net
17 income **19** interest
21 per cent **23** formula
24 monthly **27** PAYG
28 quarterly **29** balance
30 piecework

Chapter 3

StartUp assignment 3
PAGE 20

1 $0.7\dot{2}$ **2** $1085
3 1000 **4** irrational
5 2.60 **6** 35.1 L
7 $SA = 2\pi r^2 + 2\pi rh$
8 a 4.105×10^{-2} **b** 4
9 $27y^2 - 3$ **10** 360°
11 $25u^2 - 40u + 16$ **12** $\dfrac{2m}{5}$
13 $d = 3$ **14** $x = 16$
15 $(5, 0)$ **16** 24 units^2
17 a 10 units **b** $(1, 3)$ **c** $\dfrac{3}{4}$
18 $-13, -9, -5, -1$ **19** $\dfrac{-1}{3}$

20

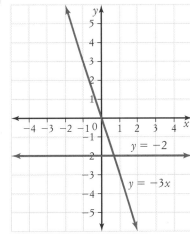

21 a one pair of opposite sides parallel
  b all sides equal
  c opposite sides equal and parallel

22 $3\sqrt{5}$

23 $y = -4$

24 Sample answer only. Other (similar) negative gradient answers are possible. Teacher to check.

25

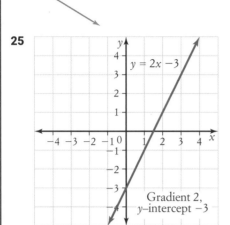

Gradient 2, y–intercept -3

26 a 2   **b** $5\frac{1}{2}$

     c 2   **d** 4

27 Yes

28 Teacher to check

Challenge: 12, circle

Writing equations of lines   PAGE 23

1 Teacher to check

2 Any equation in the form $y = 6x + c$

3 Any equation in the form $y = c$

4 Any equation in the form $y = -mx$

5 Any equation in the form $y = \frac{1}{3}x + c$

6 Any equation in the form $y = mx + 5$

7 Any equation in the form $y = \frac{1}{6}x + c$

8 Any equation in the form $y = -\frac{2}{3}x + c$

9 $x = -1$

10 Any equation in the form $y = mx + c$, with $m > 2$

11 Teacher to check

12 Teacher to check

13 Any equation in the form $y = 2x + c$

14 Any equation in the form $y = c$

15 Any equation in the form $y = mx + 5$

16 $y = -5$

17 Any equation in the form $y = mx - 5$

18 $y = -5x + 10$

19 $y = \frac{3}{10}x - 3$

20 Any equation in the form $y = mx + c$, with $m > -\frac{1}{2}$

21 $y = 3x - 7$

22 Any equation in the form $y = -\frac{3}{2}x - c$

23 $y = \frac{4}{9}x - 4$ **24** $y = -x - 4$

Linear equations code puzzle   PAGE 24

1 Y   **2** O

3 B   **4** S

5 F   **6** W

7 A   **8** L

9 R   **10** T

11 E   **12** H

13 X   **14** D

15 I   **16** J

17 N   **18** U

19 Z   **20** C

21 M   **22** G

23 V   **24** K

25 P

Answer: What did one parallel line say to another at the party?
'Excuse me, you look very familiar but we've never met, have we?'

Graphing lines crossword   PAGE 31

Across

3 steepness   **6** intercept

8 origin   **10** graph

13 line   **14** general

17 equation   **18** constant

21 reciprocal   **23** surd

24 hypotenuse   **26** negative

27 positive   **28** vertical

29 exact

Down

1 form   **2** length

4 plane   **5** Cartesian

7 gradient   **9** Pythagoras

11 run   **12** interval

15 rise   **16** midpoint

18 coordinates   **19** axes

20 parallel   **22** inclination

25 linear   **27** point

Chapter 4

StartUp assignment 4   PAGE 32

1 Example answer given. Other triangles with 2 equal sides and an included obtuse angle are possible. Teacher to check.

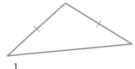

2 $\frac{1}{25k^2}$

3 a 13.25   **b** 12

4 $6ab$ **5** 115°

6 85%   **7** 3.78×10^7

8 $x = 7$

9 a 8.94 units   **b** −2

10 $\dfrac{5b}{a}$ **11** 0.8̇3̇

12 7%   **13** 10

14 a   **b**

 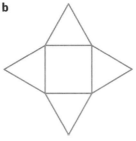

15 a 37.70 cm   **b** 113.10 cm²

16 a 384 cm²   **b** 512 cm³

17 a 10 000   **b** 1000

18 a triangular prism   **b** 5

     c rectangle, triangle   **d** 45 cm³

19 a 84.82   **b** 113.10

20 $A = \dfrac{1}{2}(a + b)h$

21 a rhombus   **b** 38.5 cm²

22 220 cm   **23** 7 cm

24 A prism has a uniform cross-section; the side faces of a pyramid meet at a point.

25 8 m

26 a 15 cm²   **b** 12 cm²   **c** 13.96 cm²

27 6 cm

Challenge: $\dfrac{2}{3}$

A page of solid shapes   PAGE 35

The first answer is the surface area, followed by the volume.

1 54 cm², 27 cm³   **2** 46 cm², 15 cm³

3 100.53 cm², 75.40 cm³   **4** 64 cm², 28 cm³

5 576 cm², 512 cm³   **6** 222 cm², 132 cm³

7 294 cm², 343 cm³   **8** 716 cm², 1280 cm³

9 150.80 cm², 100.53 cm³   **10** 472 cm², 360 cm³

11 240 cm², 180 cm³   **12** 1320 cm², 2880 cm³

13 360 cm², 400 cm³   **14** 240 cm², 264 cm³

15 113.10 cm², 113.10 cm³   **16** 678.58 cm², 1017.88 cm³

17 37.70 cm², 16.76 cm³   **18** 314.16 cm², 523.60 cm³

19 628.32 cm², 1005.31 cm³   **20** 276.46 cm², 385.37 cm³

Surface area and volume crossword (Advanced)   PAGE 37

Across

2 length   **3** square

6 perpendicular   **8** height

10 radius   **12** hemisphere

14 closed   **16** diameter

17 cone   **20** apex

22 rectangular   **23** capacity

24 open   **25** surface

26 pi   **27** net

28 triangular   **29** volume

Down

1 slant   **3** sphere

4 prism   **5** oblique

6 parallelepiped   **7** cross-section

9 trapezoidal   **11** squared

13 Pythagoras   **15** formula

16 dimension   **18** semicircle

19 composite   **21** pyramid

23 cubed

Chapter 5

StartUp assignment 5   PAGE 47

1 −4   **2** $3300, $1800

3 $m^2 + 6m$ **4** 135°

5 7.46×10^{-6} **6** 6

7 false, but a square is a rhombus

8 $\dfrac{1}{4}$ **9** 376.99 cm²

10 $x = 16\dfrac{1}{2}$ **11** 126

12

13 $45.00

14 a 23°   **b** 30 cm²

15 a 1, 3, 9   **b** 1, 3, 17, 51

16 a $21m^3$ **b** $5x$ **c** $12k^4$

     d $-8g^2$ **e** $16pq$ **f** $9u^8$

17 a 4   **b** $8y$

18 1, 4, 9, 16, 25, 36, 49, 64, 81, 100

19 a $3x + 21$ **b** $10m - 2n$ **c** $4 - 12g$

     d $-12p - 10$ **e** $m^2 + 7m$ **f** $3p^2 - 4p$

     g $6k^2 - 12k$ **h** $-15y + 5y^2$

20 a $3(x + 4)$ **b** $10(m - 2)$ **c** $p(3p - 5)$

     d $-2(4m - 9)$ **e** $-2y(7y + 4)$

Challenge: Numerous solutions.

Balance 6 coins and 6 coins to find which side holds the counterfeit. Then balance 3 coins and 3 coins from the side identified as holding the counterfeit, to again find which side holds the counterfeit. Then choose 2 coins and balance 1 and 1. If they balance, then the 3rd coin is the counterfeit. If they don't, then the lighter one is the counterfeit.

Algebraic fractions   PAGE 49

1 $\dfrac{3x}{5}$ **2** y

3 $\dfrac{7d}{4}$ **4** $\dfrac{4h}{3}$

5 $\dfrac{23h}{15}$ **6** $\dfrac{53k}{14}$

7 $\dfrac{47x}{24}$ **8** $\dfrac{5}{x}$

9 $\dfrac{32}{3d}$

10 $\dfrac{3y}{2}$

11 $2p$

12 r

13 $\dfrac{2e}{5}$

14 $\dfrac{17d}{12}$

15 $\dfrac{20p}{21}$

16 $\dfrac{-36}{5d}$

17 r

18 $\dfrac{-g}{4}$

19 $\dfrac{ey}{3}$

20 $\dfrac{8rt}{3}$

21 $45y$

22 $20g$

23 $10f$

24 $\dfrac{10hg}{3}$

25 $\dfrac{11d}{6}$

26 $\dfrac{3r^2}{5}$

27 $\dfrac{25e^2}{49d^2}$

28 $\dfrac{3d}{10}$

29 $\dfrac{5}{2}$

30 $\dfrac{21ef}{4}$

31 $\dfrac{16a}{9c}$

32 $4y^2$

33 $\dfrac{h}{2g^2}$

34 $\dfrac{21d}{10}$

35 $\dfrac{9mn}{5}$

36 $\dfrac{7h}{24f}$

Factorising puzzle (Advanced)

PAGE 50

1 D	**2** K	**3** R	**4** L	**5** T
6 X	**7** A	**8** Z	**9** Q	**10** I
11 B	**12** S	**13** U	**14** A	**15** M
16 E	**17** G	**18** T	**19** J	**20** Y
21 N	**22** O	**23** P	**24** H	**25** W
26 E	**27** S	**28** C	**29** V	**30** F

What is the difference between a doctor and an algebra student?

The doctor rectifies us while the student factorises.

Algebra crossword

PAGE 52

Across

3 product

8 binomial

10 numerator

13 brackets

14 indices

16 index

18 base

19 factors

Down

1 constant

2 quadratic

3 power

4 term

5 factorise

6 highest

7 coefficient

9 algebraic

11 reciprocal

12 HCF

15 expand

17 denominator

Chapter 6

StartUp assignment 6

PAGE 60

1 $125a^3$

2 B

3 6

4 $3 - 2y$

5 trapezoidal prism

6 a 169.65 cm^3

b 169.65 cm^2

7 $\dfrac{5}{8}$

8 $k = 10$

9 36

10 $d = 26\dfrac{2}{3}$

11 $y = 0$

12 11.5%

13 length 125 cm, width 100 cm

14 560 km

15 a 19   **b** 2   **c** 4

d 2.42   **e** 2

f

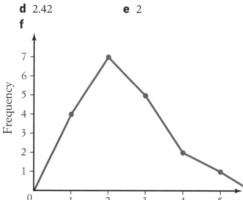

16 A census surveys the whole population; a sample surveys a part of the population.

17 a 64   **b** 24   **c** 64

d 11   **e** 64.62

18 a

Score	f	cf
1	2	2
2	5	7
3	6	13
4	8	21
5	5	26
6	2	28
7	1	29
8	1	30

b 4   **c** 4

d 7   **e** 13

Challenge: 63 matches, 6 rounds

Line of best fit

PAGE 62

3 Equation of the line of best fit should be close to $y = -1.75x + 53.42$.

Data crossword
PAGE 64

Across

2 quartiles   6 histogram
7 scatter   8 stem
11 bias   13 spread
14 order   15 five number
18 whisker   20 median
21 interquartile   22 mean
24 location   26 strong

27 bivariate   28 mode

Down

1 data   3 lower
4 plot   5 distribution
7 symmetrical   9 upper
10 range   12 skewed
15 frequency   16 middle
17 fifty   19 cumulative
23 cluster   25 outlier

Statistical calculations
PAGE 73

Data set	Mean	Mode	Median	Range	Interquartile range	Standard deviation
A	14.07	14	14	7	2	1.84
B	37.83	–	35	57	19.5	14.84
C	9.24	9	9	5	2	1.24
D	2.42	2	2	4	1	1.09
E	2.19	2	2	5	2	1.33
F	74.92	53, 71, 76, 84	76	45	21.5	13.04
G	4.75	5	5	8	3	2.33
H	9.75	10	10	8	3	2.33
I	14.25	15	15	24	9	7.00
J	6.44	5	5	20	4	5.27

Chapter 7

StartUp assignment 7
PAGE 74

1 $\frac{1}{9}$

2 $20\frac{5}{6}$ m/s

3 A

4 $\equiv$

5 $83\frac{1}{3}\%$

6 $p = 78$

7 2 km

8 a $\sqrt{41}$

b $\frac{4}{5}$

9 1 050 000

10 mode

11 $9 : 6 : 2$

12 a $144\ cm^2$

b $112\ cm^3$

13 6×10^{-3}

14 $p = 4\frac{2}{3}$

15 false

16 $y = -7$

17 a $10a - 6$

b $-12a + 24$

18 Yes

19 $k = 16$

20 $r = -3$

21 20, 18, 16, 14, 12, 10

22 $y = 6$

23 $m = 8\frac{1}{3}$

24 a $14r + 4$

b $t^2 + 8t - 9$

25 $p = 33$

26 $h = 9\frac{1}{2}$

27 37, 39, 41

28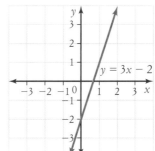

29 110°

30 $x = 3$

31 a $240

b 3 hours

32 Yes

33 $x = 7$

Challenge: 83 emus, 167 pigs

Graphing inequalities
PAGE 76

1

2

3

4

5

6

7 $-3 \quad -2 \quad -1 \quad 0 \quad 1$

8 $1 \quad 2 \quad 3 \quad 3.5 \quad 4 \quad 5 \quad 6$

9 $1 \quad 2 \quad 3 \quad 4 \quad 5$

10 -1.5
$-3 \quad -2 \quad -1 \quad 0 \quad 1 \quad 2$

11 $-2 \quad -1 \quad 0 \quad 1 \quad 2$

12 $-4 \quad -3 \quad -2 \quad -1 \quad 0 \quad 1$

13 $-2 \quad -1 \quad 0 \quad 1 \quad 2 \quad 3$

14 $2 \quad 3 \quad 4 \quad 5 \quad 6 \quad 7$

15 $0 \quad 0.5 \quad 1 \quad 1.5 \quad 2$

16 $-7 \quad -6 \quad -5 \quad -4 \quad -3 \quad -2$

17 $-3 \quad -2.5 \quad -2 \quad -1.5 \quad -1 \quad -0.5$

18 $0 \quad 1 \quad 2 \quad 3 \quad 4 \quad 5$

19 $2 \quad 4 \quad 6 \quad 8 \quad 10 \quad 12$

20 $-5 \quad -4 \quad -3 \quad -2 \quad -1 \quad 0$

21 $-1 \quad -\frac{1}{2} \quad 0 \quad \frac{1}{2} \quad 1 \quad 1\frac{1}{2}$

22 $5 \quad 5\frac{1}{2} \quad 6 \quad 6\frac{1}{2} \quad 7 \quad 7\frac{1}{2}$

23 $-7 \quad -6 \quad -5 \quad -4 \quad -3 \quad -2 \quad -1$

24 $2 \quad 2\frac{1}{2} \quad 3 \quad 3\frac{1}{2} \quad 4 \quad 4\frac{1}{2}$

Logarithms review
PAGE 78

1 a 3   **b** 8
  c 3   **d** 2
  e -1 **f** -2
  g $\frac{1}{2}$ **h** 0

2 a $\log_4 64 = 3$ **b** $\log_5 25 = 2$
  c $\log_2 32 = r$ **d** $\log_y z = 6$

3 a $10^2 = 100$ **b** $2^3 = 8$
  c $3^7 = b$ **d** $a^2 = 16$

4 a 5   **b** 3
  c 2   **d** 3

e 1   **f** 3
g $\frac{1}{2}$ **h** 6

5 a $\log_a 30$ **b** $\log_a 5$
  c $\log_a 12$ **d** $\log_a 4 + 2$
  e 0   **f** $\log_a 3 - 2$
  g $\log_a 4 + \frac{1}{2}$ **h** $\log_a\left(\dfrac{3c\sqrt[3]{d}}{e^2}\right)$

6 a $\log_a x + \log_a y$ **b** $n\log_a x$
  c $\log_a p + \log_a q - \log_a r$ **d** $\log_a 4 + 2\log_a 5$
  e $\frac{1}{2}\log_a x + \frac{1}{2}\log_a y$ **f** $\log_a c - 2\log_a d$

7 a $x + z$ **b** $\frac{1}{2}x$
  c $2x + z$ **d** $y + z - x$

8 a 1.5564   **b** -0.7782
  c 1.7782   **d** -0.2218
  e 2.5564   **f** 1.8891

9 a $x = 6$ **b** $x = 8$
  c $x \approx 2.97$ **d** $x \approx 9.67$
  e $x \approx 8.48$ **f** $x \approx -1.19$

10 a $x = 81$ **b** $x = 3125$
  c $x = 2$ **d** $x = 6$
  e $x = 10$ **f** $x = 4^x$

Equations and inequalities crossword
PAGE 80

Down

  1 factorise   **2** less
  5 brackets   **9** line
10 solution   **11** negative
14 two   **15** quadratic
18 LCM   **21** formula

Across

  3 subject   **4** substitute
  6 expand   **7** greater
  8 inequality   **12** solve
13 equation   **16** root
17 variable   **19** inverse
20 multiple   **21** fraction
22 RHS   **23** check
24 LHS

Changing the subject of a formula
PAGE 90

1 $r = \dfrac{I}{Pn}$ **2** $w = \dfrac{V}{lh}$

3 $t = \dfrac{d}{s}$ **4** $r = \dfrac{C}{2\pi}$

5 $r = \dfrac{360l}{2\pi\theta}$ **6** $r = \sqrt{\dfrac{A}{\pi}}$

7 $A = \dfrac{V}{h}$ **8** $x = \dfrac{2A}{y}$

9 $h = \dfrac{V}{\pi r^2}$ **10** $s = \sqrt{A}$

11 $b = \dfrac{2A}{h}$

12 $h = \dfrac{2A}{a + b}$

13 $r = \sqrt{\dfrac{360A}{\theta\pi}}$

14 $r = \sqrt{\dfrac{v}{\pi h}}$

15 $w = \dfrac{A}{l}$

16 $c = y - mx$

17 $d = \dfrac{C}{\pi}$

18 $P = \dfrac{A}{(1 + r)^n}$

19 $r = \sqrt[3]{\dfrac{3V}{4\pi}}$

20 $n = \dfrac{A}{180} + 2$

21 $m = \dfrac{v - c}{x}$

22 $r = \sqrt{\dfrac{SA}{4\pi}}$

23 $x = \dfrac{-by - c}{a}$

24 $y = \dfrac{-ax - c}{b}$

25 $a = \dfrac{2A}{h} - b$

26 $b = \sqrt{c^2 - a^2}$

27 $l = \dfrac{SA - \pi r^2}{\pi r}$

28 $x = \dfrac{y - y_1}{m} + x_1$

29 $h = \dfrac{SA - 2\pi r^2}{2\pi r}$

30 $r = \sqrt[n]{\dfrac{A}{P}} - 1$

Chapter 8

StartUp assignment 8   PAGE 91

1 1.39

2 $\dfrac{1}{8}$

3 40 000

4 198

5 143.35

6 $5y$

7 3770 cm³

8 6150

9 360°

10 108

11 $\dfrac{3}{10}$

12 56°

13 $1300, $650, $2600

14 0

15 $456.50

16 a $-8, -2, 1, 7$

   b

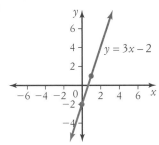

   c 3   **d** -2 **e** $\dfrac{2}{3}$

17 $0, -4, -3, 5$

18 a $x = -2$ **b** $y = 18$

19 Example answer only.

    Other similar answers possible.

    Teacher to check.

20 a 12   **b** 48   **c** 3

21 a $y = 4x + 3$ **b** $y = x^2$

22 a 1   **b** $\dfrac{1}{2}$ **c** $y = \dfrac{1}{2}x + 1$

23 a 60 km/h   **b** 12:30 p.m.   **c** 150 km

24 a $-1\dfrac{1}{3}$ **b** $\dfrac{2}{5}$ **c** 8

Challenge: 16 m

Matching parabolas   PAGE 93

1 D   **2** A

3 I   **4** E

5 H   **6** L

7 B   **8** P

9 G   **10** C

11 F   **12** N

13 M   **14** K

15 J   **16** O

Graphing curves crossword   PAGE 96

Across

3 line   **6** asymptote

7 graph   **9** point

10 circle   **11** centre

13 direct   **18** coordinates

19 vertex   **22** constant

24 variable   **25** horizontal

26 axis

Chapter 9

StartUp assignment 9   PAGE 103

1 $250

2 8

3 1 h 40 min

4 $\dfrac{8}{11}$

5 48 m

6 $14 + 6\sqrt{5}$

7 201.84 cm²

8 $9889.05

9 $-\dfrac{3}{4}$

10 $(4y + 3)(y - 5)$

11 $d = 32$

12 $k \le \dfrac{1}{3}$

13 4.5

14 $y = -3$

15 $y = -\dfrac{1}{3}x$

16 $\dfrac{a}{b}$

17 a 1.1276   **b** 5.7671   **c** 1

   d 0.3420   **e** 0.3420

18 a same   **b** 40°

19 a $\dfrac{\sqrt{3}}{2}$ **b** $\dfrac{1}{\sqrt{3}}$

20 a 30° 58′   **b** 56° 26′
     c 60°   **d** 61° 56′

21

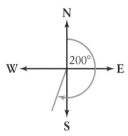

22 a 15.43   **b** 26.46
23 a 86.24   **b** 6.47
     c 15.17   **d** 2.30
24 502.51 m
25 20°
26 a 10 km   **b** 323° 8′
Challenge: $\theta = 45°$

Finding an unknown angle   PAGE 106

1 33°   **2** 68°
3 58°   **4** 80°
5 115°   **6** 124°
7 26°   **8** 66°
9 17°   **10** 27°
11 136°   **12** 137°
13 83°   **14** 63°
15 125°   **16** 117°
17 90°   **18** 90°
19 60°   **20** 70°
21 88°

Trigonometry crossword (Advanced)   PAGE 108

Across

6 second   **9** negative
10 surd   **12** theta
14 obtuse   **16** minutes
18 supplementary   **23** bearing
25 one   **26** Pythagoras
27 west   **30** WNW
33 elevation   **34** degree
36 north-west   **38** depression
39 south-west   **40** alternate

Down

1 ESE   **2** included
3 half   **4** south
5 sine   **7** cosine
8 north   **11** complementary
13 tangent   **15** ENE
17 three   **19** trigonometry
20 SSE   **21** positive
22 NNW   **24** north-east
28 hypotenuse   **29** south-east
31 opposite   **32** NNE
35 east   **37** area

Chapter 10

StartUp assignment 10   PAGE 119

1 12.2   **2** $3280.77
3 $n + 1$
4 area of a rhombus or kite
5 $9.1^2 = 3.5^2 + 8.4^2$
6 a 64   **b** $\dfrac{1}{9}$
7 a $2x + 18$ **b** $\dfrac{p^2}{5}$
8 6.28 cm²   **9** 0.000 028
10 $x = 2\dfrac{1}{2}$ **11** SSS, SAS, AAS, RHS
12 $x = 19$ **13** 502.65 cm²
14 a $-40g + 32$ **b** $13y - 30$
     c $-18a - 6$ **d** $75b - 67$
15 a $p = -5$ **b** $f = 8.5$
     c $a = -1.5$ **d** $g = 19.2$
     e $x = 13$ **f** $a = 3\dfrac{2}{3}$
16 a $y = 29$ **b** $y = 44$ **c** $y = 7$
17 a $a = -26$ **b** $a = 90$ **c** $a = 35$
18 a yes   **b** yes   **c** no

19

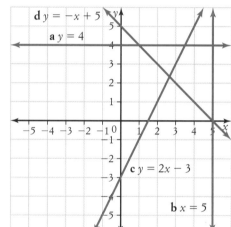

20 74 and 76

21 33, 35, 37

Challenge: Timmy

Intersection of lines   PAGE 121

1 $(3, 5)$ **2** $(1, 0)$
3 $(4, -2)$ **4** $(-2, 6)$

Simultaneous equations crossword   PAGE 122

Across

7 solve   **11** intersection   **12** point
14 satisfy   **15** substitution   **16** method
17 solution

Down

1 simultaneous   **2** problem   **3** variable

4 algebraic   **5** axes   **6** graphical

8 equations   **9** elimination   **10** linear

13 coefficient

Chapter 11

StartUp assignment 11   PAGE 128

1 0.707   **2** $\dfrac{1}{8}$

3 10 000 000   **4** 198

5 Teacher to check. The equation must be of the form $y = mx + c$ where $m < -1$ and $c > 0$.

6 $5y$ **7** 86.32 cm²

8 6150   **9** rectangle

10 108   **11** $\dfrac{3}{10}$

12 34°   **13** $1300, $650, $2600

14 360°   **15** $674.66

16 a 3   **b** -2

17 a 14   **b** 22   **c** -5 **d** 0

18

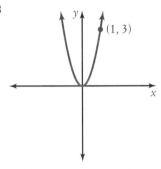

19 a 36   **b** 8   **c** 120

20 a $x = -2$ or 7   **b** $y = -1$ or 5

21 $x^2 - 6x + 9$ **22** $(x + 5)(3x - 4)$

23 a 0   **b** -15 **c** 28

24

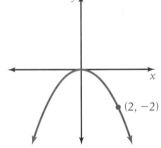

25 a 1   **b** -2

    c $\dfrac{1}{2}$ **d** $y = \dfrac{1}{2}x + 1$

26 a $-\dfrac{1}{5}$

27 a 2.83   **b** 5

Challenge: 2

Quadratic equations puzzle   PAGE 130

1 W	**2** G	**3** Q
4 X	**5** O	**6** C

7 E	**8** V	**9** H
10 D	**11** M	**12** P
13 A	**14** F	**15** U
16 T	**17** J	**18** I
19 S	**20** B	**21** O
22 E	**23** A	**24** T
25 I	**26** Y	**27** K
28 R	**29** L	**30** N

Playing this game is like graphing lines and curves on a number plane. You've got to look for the intercepts.

Quadratic equations crossword   PAGE 132

Across

2 simultaneous   **4** factorise

5 coordinates   **9** constant

11 concave   **14** symmetry

16 formula   **17** parabola

19 intercept   **21** solution

23 intersection

Down

1 coefficient   **3** quadratic

6 monic   **7** root

8 square   **9** complete

10 vertex   **12** equation

13 hyperbola   **14** substitution

15 surd   **18** axis

20 exact   **22** circle

Chapter 12

StartUp assignment 12   PAGE 139

1 1 000 000   **2** $\dfrac{11a}{15}$

3 Equal or bisect each other   **4** $80 and $32

5 376.33 cm³   **6** $916

7 $y = 36$ **8** 3

9 70   **10** $d = 7$

11 $24.50   **12** $10 200

13 30

14 a 2.4 m   **b** 34.56 m²

15 58%

16 a $\dfrac{7}{20}$ **b** $\dfrac{1}{5}$ **c** $\dfrac{1}{4}$

17 a $\dfrac{1}{6}$ **b** $\dfrac{17}{24}$ **c** $\dfrac{5}{6}$

18 $\dfrac{1}{6}$ **19** Teacher to check

20 a 0.55   **b** $\dfrac{1}{6}$

21 Impossible, it cannot happen

22 0.5

23 Reds win, Blues win, or a draw

24 $\frac{2}{5}$

25 0.02   **26** $\frac{5}{8}$

27 55%

28 a $\frac{1}{4}$ **b** $\frac{1}{13}$

   c $\frac{3}{13}$ **d** $\frac{5}{13}$

29 19.2%

Challenge: 120

Probability crossword   PAGE 141

Across

3	table	**6**	without
7	replacement	**8**	conditional
9	mutually	**13**	Venn
14	outcome	**18**	theoretical
20	diagram	**21**	event
22	die	**25**	tree
26	dependent	**27**	step
28	list	**29**	probability

Down

1	frequency	**2**	overlapping
4	compound	**5**	random
10	two way	**11**	independent
12	relative	**15**	complementary
16	experimental	**17**	dice
18	trial	**19**	expected
23	sample	**24**	exclusive

Tree diagrams   PAGE 142

1 a All possible arrangements of boys and girls.

   b i $\frac{1}{4}$ **ii** $\frac{1}{4}$ **iii** $\frac{1}{2}$

2 a PP, PF, FP, FF   **b i** $\frac{1}{4}$ **ii** $\frac{1}{4}$

3 a

   b $\frac{3}{8}$

4 $\frac{3}{4}$

5 a

		1st Die					
		1	**2**	**3**	**4**	**5**	**6**
2nd die	**1**	0	1	2	3	4	5
	2	1	0	1	2	3	4
	3	2	1	0	1	2	3
	4	3	2	1	0	1	2
	5	4	3	2	1	0	1
	6	5	4	3	2	1	0

   b i $\frac{1}{6}$ **ii** $\frac{1}{6}$ **iii** $\frac{1}{3}$ **iv** $\frac{1}{6}$

6 a 1T, 2T, 3T, 4T, 1H, 2H, 3H, 4H

   b i $\frac{1}{8}$ **ii** $\frac{1}{4}$

7 $\frac{7}{27}$ **8** B

9 a $\frac{1}{3}$ **b** $\frac{2}{3}$ **c** $\frac{1}{2}$

Two-way tables   PAGE 144

1 a 24   **b** 8

   c i $\frac{1}{8}$ **ii** $\frac{1}{6}$ **iii** $\frac{2}{3}$

2 a 70   **b** 50

   c i $\frac{43}{70}$ **ii** $\frac{9}{70}$ **d** $\frac{9}{25}$

3 a 50   **b** sport

   c i $\frac{22}{50} = \frac{11}{25}$ **ii** $\frac{20}{50} = \frac{2}{5}$ **iii** $\frac{1}{10}$

   iv $\frac{15}{50} = \frac{3}{10}$

4 a 32   **b** 16

   c i $\frac{28}{60} = \frac{7}{15}$ **ii** $\frac{44}{60} = \frac{11}{15}$ **iii** $\frac{7}{60}$

5 a 130   **b** males

   c i $\frac{9}{130}$ **ii** $\frac{37}{130}$ **iii** $\frac{116}{130} = \frac{58}{65}$

   d It may be too far to walk.

Chapter 13

StartUp assignment 13   PAGE 150

1 a 4.20   **b** $21 386.70   **c** 678.58 m³

2 The mode is the most common value.

3 1 : 5000   **4** $b + 3$

5 285°   **6** $x = 3\frac{1}{2}$

7 $x = 44$

8 10.5 km

9 121.5 cm²

10 24°

11 17 units

12 $262.61

13 $\dfrac{1}{2}$

14 133°

15 Example answer given. Other rhombuses are possible. Teacher to check.

16 $\angle AED$

17 360°

18 equilateral

19 a 275 km   **b** 6.5 cm

   c 1 : 5 000 000

20 Example answer given. Other obtuse-angled triangles are possible. Teacher to check.

21 a $a = 72$ **b** $y = 110$ **c** $p = 240$

    d $e = 60$ **e** $u = 45$ **f** $x = 135$

22 8

23 a false   **b** true   **c** false

24 25°, 25°

25 a A and B   **b** SAS and RHS

26 a $\angle J$ **b** $d = 32\dfrac{1}{2}$

Challenge:

Other solutions possible.

Proving properties of quadrilaterals   PAGE 152

1 a SSS

  b Teacher to check matching angles

  c axis, bisects

  d Teacher to check diagonal

  e angles in an isosceles triangle

  f AAS

  g $\angle AEB$

  h 90°

  i right angles

2 a AAS

  b $\angle YXW$

  c Teacher to check matching angles

  d equal, equal

  e Teacher to check diagonal

  f vertically opposite angles

  g AAS

  h Teacher to check matching sides

  i bisect

3 a SSS

  b Teacher to check matching angles

  c 90°

  d right angles, bisect

4 a SAS   **b** QS   **c** equal

5 a They are equal and bisect each other

  b SSS

  c base angles of an isosceles triangle

  d Teacher to check matching angles

  e 45°

  f 90°

  g equal, right angles, bisect

Geometry crossword   PAGE 154

Across

  3 ratio

  9 similar

 16 bisect

 25 enlargement

 28 exterior

 30 dodecagon

 32 reduction

  8 nonagon

 12 quadrilateral

 20 matching

 26 SSS

 29 symmetry

 31 interior

 33 polygon

Down

  1 pentagon

  3 RHS

  5 right

  7 diagonal

 11 trapezium

 14 obtuse

 17 triangle

 19 isosceles

 22 rotational

 24 geometry

 27 SAS or SSS

  2 rhombus

  4 test

  6 AAS

 10 parallelogram

 13 equilateral

 15 decagon

 18 eighty

 21 congruent

 23 rectangle

 26 square

Complete the proofs   PAGE 161

1 a $\angle KML = \dfrac{180° - 90°}{2} = 45°$ (angle sum of isosceles $\triangle KML$)

  b $\therefore \angle PMN = 180° - 45° = 135°$

  c $\therefore \angle N = \dfrac{180° - 135°}{2} = 22.5°$ (angle sum of isosceles $\triangle PNM$)

2 a (alternate angles, $AC \parallel DG$)

  b $\therefore \angle ABF = x + y$

  c $\therefore \angle BFG = \angle ABF$ (alternate angles, $AC \parallel DG$)

  d $\therefore \angle BFG = x + y$

  e $\therefore \angle BFG = \angle BEF + \angle EBF$

3 a In $\triangle DAB$ and $\triangle CAB$

  b $DA = AC$ (equal radii)

  c $DB = BC$ (equal radii)

  d AB is common

e ∴ △*DAB* ≡ △*CAB* (SSS)

f ∴ ∠*DAB* = ∠*CAB* (matching angles in congruent triangles)

g ∴ *BA* bisects ∠*DAC*

4 a ∴ ∠*PWN* = 180° − 90° − θ = 90° − θ (angle sum of △*WNP*)

b Also ∠*TWN* = 90° − (90° − θ) = θ

c ∴ ∠*TWN* = ∠*P* = θ

d In △*PWN* and △*WTN*

e ∠*WNP* = ∠*WNT* = 90° (given)

f ∠*P* = ∠*TWN* = θ (proven above)

g ∴ △*PWN* ||| △*WTN* (AA)

h ∴ $\frac{WN}{9} = \frac{4}{WN}$ (matching sides in similar triangles)

i $WN^2 = 36$

j *WN* = 6

Chapter 14

StartUp assignment 14 PAGE 162

1 3.34

2

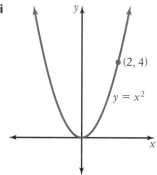

3 a $9p^2 - 64$ **b** $21 - 8\sqrt{5}$

4 6 200 000

5 $x = \dfrac{-b \pm \sqrt{b^2 - 4ac}}{2a}$

6 2000 **7** $SA = 3\pi r^2$

8 40° **9** pentagon

10 $x \le 2\dfrac{1}{13}$

11 $\dfrac{1}{36}$

12 a (3, −2) **b** $4\sqrt{2}$ units

13 circle

14 a i

[Graph: $y = x^2$, point (2, 4)]

ii

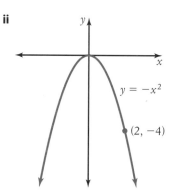

[Graph: $y = -x^2$, point (2, −4)]

b One is concave up, the other concave down

15 a i

[Graph: $y = 3x^2$, point (2, 12)]

ii

[Graph: $y = 7x^2$, point (2, 28)]

iii

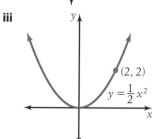

[Graph: $y = \frac{1}{2}x^2$, point (2, 2)]

b The steepness of the graph

16 a

[Graph: $y = \frac{1}{x}$, points (1, 1) and (−1, −1)]

b

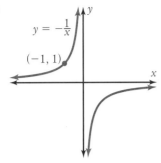

$y = -\frac{1}{x}$

$(-1, 1)$

c

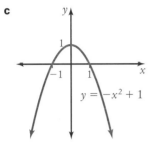

$y = -x^2 + 1$

d

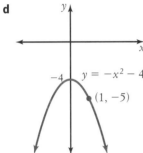

$y = -x^2 - 4$

$(1, -5)$

e

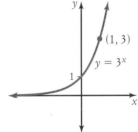

$(1, 3)$

$y = 3^x$

f

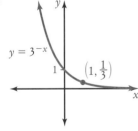

$y = 3^{-x}$

$\left(1, \frac{1}{3}\right)$

17 a $x = \dfrac{y - 4}{3}$ **b** $x = \pm\sqrt{\dfrac{y}{5}}$

c $x = -\dfrac{2}{y}$

18 a -2 **b** -5

c -5 **d** $a^2 + 2a - 5$

e $16a^2 + 8a - 5$ **f** $9a^2 - 6a - 5$

19 a

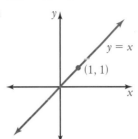

$y = x$

$(1, 1)$

b

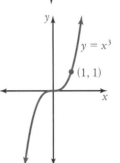

$y = x^3$

$(1, 1)$

c

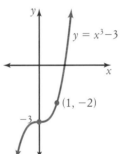

$y = x^3 - 3$

$(1, -2)$

-3

Challenge: $x = 13$

1

	$f(x) = 3 - x$	$f(x) = x^2 + 6x$	$f(x) = \dfrac{1}{3x - 1}$	$f(t) = t^2 - \dfrac{t}{4}$	$f(t) = 2t^2 - 3t + 5$
$f(1)$	2	7	$\dfrac{1}{2}$	$\dfrac{3}{4}$	4
$f(-2)$	5	-8	$-\dfrac{1}{7}$	$4\dfrac{1}{2}$	19
$f(0)$	3	0	-1	0	5
$f(5)$	-2	55	$\dfrac{1}{14}$	$23\dfrac{3}{4}$	40
$f\left(2\dfrac{1}{2}\right)$	$\dfrac{1}{2}$	$21\dfrac{1}{4}$	$\dfrac{2}{13}$	$5\dfrac{5}{8}$	10
$f(c)$	$3 - c$	$c^2 + 6c$	$-\dfrac{1}{30 + 5}$	$c^2 + \dfrac{c}{4}$	$2c^2 - 3c + 5$
$f(-c)$	$3 + c$	$c^2 - 6c$	$-\dfrac{1}{30 + 5}$	$c^2 + \dfrac{c}{4}$	$2c^2 + 3c + 5$
$f(c + 2)$	$1 - c$	$c^2 + 10c + 16$	$\dfrac{1}{3c + 5}$	$\dfrac{4c^2 + 15c + 14}{4}$	$2c^2 + 5c + 7$
$f(3 - c)$	c	$c^2 - 12c + 27$	$\dfrac{1}{8 - 3c}$	$\dfrac{4c^2 - 23c + 33}{4}$	$2c^2 - 9c + 14$
$f(p^2)$	$3 - p^2$	$p^4 + 6p^2$	$\dfrac{1}{3p^2 - 1}$	$p^4 - \dfrac{p^2}{4}$	$2p^4 - 3p^2 + 5$

2 a 22   **b** $\sqrt{31}$ **c** 16   **d** -1

3 a 1   **b** ±3 **c** 2   **d** 5

Graphing translations of functions   PAGE 165

1

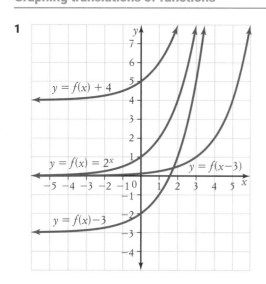

2

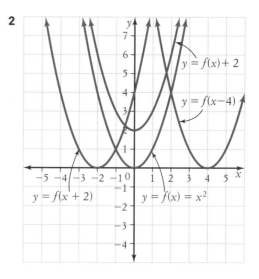

3

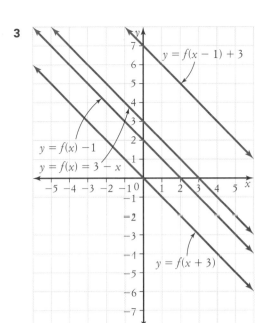

4

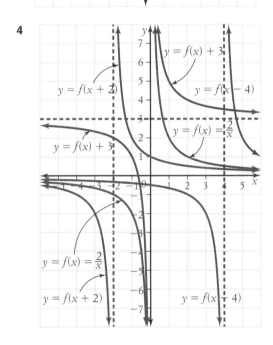

Chapter 15

StartUp assignment 15 PAGE 170

1 $\dfrac{1}{2}$

2 $-\sqrt{3}$

3 SSS, SAS, AA, RHS

4 $r = -2\dfrac{4}{7}$

5 a 251.33 cm²   **b** 3.75   **c** 38.26 cm²

6 1.6%

7 $\dfrac{d^3}{8c^6}$

8

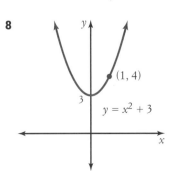

9 $864

10 $x = 4, y = -1$

11 $x - 3y + 9 = 0$

12 $9 : 25$

13 $24 112

14

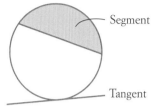

15 a $a = 78$ **b** $w = 154$

    c $m = 35, n = 145$ **d** $p = 65, r = 85$

16 a SSS or SAS

    b In $\triangle OCA$ and $\triangle OCB$, $OA = OB$

       (equal radii)

       OC is common

       $AC = CB$ (given)

       $\therefore \triangle OCA \equiv \triangle OCB$ (SSS)

    c $\angle OCA = \angle OCB = 90°$ (matching angles in congruent

       triangles)

       $\therefore OC \perp AB$

17 17 cm   **18** 3.85 cm²

19 a Equal angles in isosceles $\triangle OAB$

    b $2x + 2y = 180°$

    c $\angle ABC = x + y = 90°$

20 a $x = 5\dfrac{3}{5}$ **b** $d = 28$

    c $y = \pm 12$ **d** $n = -9$ or 4

21 a SAS

    b In $\triangle AOB$ and $\triangle DOC$, $OA = OD$ (equal radii)

       $OB = OC$ (equal radii)

       $\angle AOB = \angle DOC$ (vertically opposite angles)

       $\therefore \triangle AOB \equiv \triangle DOC$ (SAS)

    c $AB = DC$ (matching sides in congruent triangles)

    d Matching angles in congruent triangles, equal angles

       in isosceles triangles

    e Alternate angles equal

Challenge: $\dfrac{1}{2}$

Circle geometry review PAGE 173

1 $a = 65, b = 90$ **2** $u = 52$

3 $k = 49$ **4** $x = 65, y = 25$

5 $m = 72$ **6** $d = 105, e = 105$

7 $p = 14$ **8** $h = 6$

9 $OP = 69$ **10** $v = 220,$

                         $w = 70, x = 110$

11 $d = 102$ **12** $c = 45$

13 $n = 15, y = 7$ **14** $a = 60, b = 60, c = 30$

15 $v = 3\dfrac{1}{3}$ **16** $w = 103$

17 $a = 46$ **18** $r = 30, p = 50$

19 $x = 5, y = z = 38$ **20** $d = 60, e = 30$

21 $t = 50$ **22** $k = 5$

23 $x = 3$ **24** $r = 2$

HOMEWORK ANSWERS

CHAPTER 1

Surds 1
PAGE 6

Part A

1 a 24, 32 **b** 30

2 $m = 5\frac{3}{4}$ **3** $x = 110$

4 (0, 0), 6 units **5** 256 cm^2

6 (1, 7) **7** discrete

Part B

1 49, 100

2 a $-18p - 24$ **b** $x^2 - 10x + 21$

3 a x **b** 5

4 $\sqrt{88}$ $\sqrt{125}$

5 a irrational **b** rational

Part C

1 a $4\sqrt{10}$ **b** $\frac{\sqrt{10}}{10}$

c $5\sqrt{3}$ **d** $10\sqrt{6} - 5\sqrt{5}$

2 a $3\sqrt{3} + \sqrt{2}$ **b** $-5\sqrt{5}$

Part D

1 a false **b** true

c true **d** false

2 a I **b** R **c** R **d** I

Surds 2
PAGE 8

Part A

1 24 cm^2 **2** $\frac{3}{2}$

3 6 : 8 : 3 **4** $x^2 - 8x + 3$

5 $(7x - 3)^2$ **6** {HH, HT, TH, TT}

Part B

1 a $5\sqrt{11}$ **b** $28\sqrt{3}$

2 a $13\sqrt{2}$ **b** $5\sqrt{2} - 7\sqrt{5}$ **c** $23\sqrt{5} - 9\sqrt{7}$

Part C

1 a $15\sqrt{30}$ **b** 8 **c** -1

2 a $\frac{5\sqrt{2}}{2}$ **b** $\frac{2\sqrt{7}}{3}$ **c** $\frac{5\sqrt{3} - 3}{6}$

Part D

1a Multiply both the numerator and denominator by $\sqrt{5}$.

b The $\sqrt{5}$ in the denominator becomes rational (5) when multiplied by itself.

2 a $\sqrt{x} \times \sqrt{y}$ **b** $\frac{\sqrt{x}}{\sqrt{y}}$

c $a^2 - 2ab + b^2$

3 a $9 - 4\sqrt{5}$ **b** $-10 + 21\sqrt{2}$

CHAPTER 2

Compound interest
PAGE 18

Part A

1 $200 **2** 8 and 9

3 $\sqrt{82}$

4 a $SA = 2\pi r^2 + 2\pi rh$ **b** $V = \pi r^2 h$

5 a 4 **b** 3 **c** 3

Part B

1 a 0.175 **b** 0.05

2 $4612.50

3 a 52 **b** 12 **c** 365

4 a $63 814.08 **b** $68 106.54

Part C

1 $552.08 **2** 6.5%

3 $20 058.66

4 a $23 936.29 **b** $3 936.29

5 $581.46

Part D

1 a simple interest **b** compound interest

c the number of periods **d** principal

2 a $3325 **b** $41 325

3 a $16 766.12 **b** $266.12

CHAPTER 3

Coordinate geometry
PAGE 26

Part A

1 $500 **2** 15%

3 a 1000 **b** 0.24 or $\frac{6}{25}$

4 $n = 65$ (alternate angles on parallel lines), $m = 25$ (angles in a right angle)

Part B

1 a

x	-2	-1	0	1
y	-4	-3	-2	-1

b

x	-1	0	1	2
y	-7	-3	1	5

2 a neither **b** positive

c negative **d** neither

Part C

1 a $\sqrt{20}$ or $2\sqrt{5}$ **b** (4, 2) **c** $\frac{1}{2}$

2 a $y = -3x + 6$

x-intercept 2, y-intercept 6.

3 -3 **4** It does.

Part D

1 a

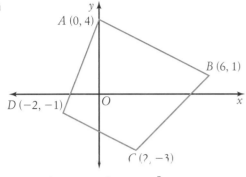

b $m_{AB} = -\dfrac{1}{2}, m_{CD} = -\dfrac{1}{2}, m_{AD} = \dfrac{5}{2}, m_{BC} = 1$

c trapezium

2 $y = -x - 2$ **3** $x = 3$

Graphing lines PAGE 28

Part A

1 a $7m(7m - 5n)$ **b** $(x + 2)(8 - p)$

2 a $14x^4y^2$ **b** $\dfrac{16m^4}{9}$

3 a 69   **b** 30s and 40s

4 $W = 21$

Part B

1 a 2   **b** $-\dfrac{1}{2}$

2 $y = \dfrac{1}{2}x + 7$

3

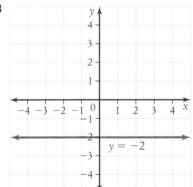

4 a $m_1 = 4, m_2 = 4$: parallel because $m_1 = m_2$

  b $m_1 = -\dfrac{1}{2}, m_2 = 2$: perpendicular because $m_1 = \dfrac{-1}{m_2}$

Part C

1 $y = -3x - 3$

2 a $y = 5x + 3$ **b** $y = x - 1$

3 a $y = -\dfrac{1}{5}x + 3$ **b** $y = \dfrac{1}{3}x - 7$

Part D

1 a parallel   **b** perpendicular

2 a $\dfrac{3}{2}$ **b** $(1, -1)$

  c $m = -\dfrac{2}{3}$ **d** $y = -\dfrac{2}{3}x - \dfrac{1}{3}$ or $y = \dfrac{-2x - 1}{3}$

3 a $y = \dfrac{1}{2}x + 2$ **b** $y = -2x - 3$

CHAPTER 4

Surface area PAGE 38

Part A

1 $x = 7$ **2** $37.50

3 a $\dfrac{56}{65}$ **b** $\dfrac{56}{65}$ **c** $\dfrac{33}{56}$

4 a Monday   **b** Wednesday

5 2.5 cm

Part B

1 a 384 units^2 **b** 336 units^2

  c 706.86 units^2 **d** 16 cm^2

  e 326.73 cm^2

2 a $x = 12$ **b** $y = 3.9$

Part C

1 a 396 cm^2 **b** 792 cm^2

2 a 151.33 m^2 **b** 251.33 m^2

Part D

1 The total area of all the faces of the solid.

2 cross-section, polygon

3 $2\pi rh$

4 a 1894.5 m^2 **b** 14.4 m^2

Surface area 2 PAGE 41

Part A

1 468

2 a $\dfrac{1}{16a^4}$ **b** $343r^3$

3 $2\sqrt{3}$

4 a $(3, 9)$ **b** $4\sqrt{10}$ units   **c** $\dfrac{1}{3}$

5 $\dfrac{1}{12}$

Part B

1 a 512 cm^2 **b** 360 m^2

2 28.8 m^2 **3** 52 m^2

Part C

1 659 cm^2 **2** 1883 mm^2

3 1662 m^2 **4** 660 cm^2

Part D

1 a pyramid   **b** apex

2 The slant height of a cone is the height from its apex to the base along a side face, while the perpendicular height is the 'vertical' distance from the apex, at right angles to the base.

3 A hemisphere is half a sphere. Its surface area is made up of half the surface area of a whole sphere plus the circular base.

$$\begin{aligned} \text{Surface area} &= \frac{1}{2} \times 4\pi r^2 + \pi r^2 \\ &= 2\pi r^2 + \pi r^2 \\ &= 3\pi r^2 \end{aligned}$$

4 a 723.60 cm^2 **b** 357.36 m^2

Part A

1 $-3x^2 + 19x - 6$ **2** 1.75

3 0.004 09   **4** 60 cm²

5 a

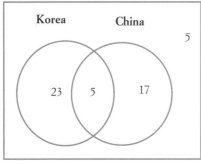

  b 5   **c** 23

  d $\dfrac{14}{25}$

Part B

1 7776 cm²   **2** 1728 cm²

3 3317.5 m²   **4** 25 699.6 mm²

Part C

1 849 cm³   **2** 5429 mm³

3 6371 m³   **4** 1257 cm³

Part D

1 Finds the volume of a pyramid, where A is the area of the base and h is the perpendicular height.

2 $V = \dfrac{4}{3}\pi r^3$

3 a 320.64 m³   **b** 66 065.08 mm³

  c 540 m³

CHAPTER 5

Index laws 2   PAGE 54

1 $\dfrac{1}{49}$ **2** 10 cm by 20 cm

3 Positively skewed   **4** 200 cm²

5 x-intercept $\dfrac{1}{2}$, y-intercept 2

6 $\dfrac{3}{4}$

Part B

1 $\dfrac{9}{4}$ **2** $\dfrac{125}{343}$

3 $\dfrac{9p^8}{25d^6}$ **4** $\dfrac{3m^2}{p^2}$

5 $\dfrac{1}{16h^2}$ **6** $1\dfrac{1}{2}$

7 2   **8** 27

Part C

1 a $(100x)^{\frac{1}{10}}$ **b** $(2mn)^{\frac{1}{3}}$

  c $y^{\frac{5}{3}}$ **d** $p^{-\frac{5}{4}}$

2 a $4ab^3$ **b** $7x^{\frac{2}{3}}$

  c $512x^6y^{15}$ **d** $\dfrac{1}{100m^2n^4}$

Part D

1 a the cube root of the number ($\sqrt[3]{}$)

  b the 6th root of the number ($\sqrt[6]{}$)

2 a $\sqrt[n]{a^m}$ **b** $\left(\sqrt[n]{a}\right)^m$

3 $\dfrac{1}{216}$

4 a $625d^{20}$ **b** $-4n^4$

Algebraic fractions   PAGE 56

Part A

1 $21a^3$ **2** $100x^2 - 4y^2$

3 $y = -2x + 3$

4 a $\dfrac{1}{10}$ **b** $\dfrac{1}{2}$ **c** $\dfrac{2}{5}$

5 $\dfrac{3}{4}$

Part B

1 a $\dfrac{5}{21}$ **b** $\dfrac{23}{18}$ **c** $\dfrac{15}{28}$ **d** $\dfrac{2}{3}$

2 a $\dfrac{mn^3}{6}$ **b** $\dfrac{2}{t}$ **c** $25x^6y^4$ **d** $81a^4b^8$

Part C

1 $\dfrac{17x}{30}$ **2** $\dfrac{y}{48}$ **3** $\dfrac{2m}{35}$ **4** $\dfrac{13y}{20}$

5 $\dfrac{12y^2}{x^2}$ **6** $\dfrac{m^2}{12}$ **7** $\dfrac{3p}{35a}$ **8** $\dfrac{4}{15n}$

Part D

1 a denominator, numerators

  b numerators, denominators

  c reciprocal

2 6

3 a 3^{-2} is the same as $\dfrac{1}{3^2}$, so the final answer is $\dfrac{1}{9}$

  b Any number to the power of 0 is 1, so $3^0 = 1$

Special binomial products   PAGE 58

Part A

1 -27 **2** Pythagorean triad

3 -4 **4** $0.1\dot{6}$

5 155°   **6** $14a + 10b$

7 $m = 68, y = 112$

Part B

1 a $p^2 + 3p - 4$ **b** $y^2 - 12y + 27$

2 a $(x - 3)(x + 11)$ **b** $(a - 4)(a - 7)$

Part C

1 a $x^2 - 16x + 64$ **b** $4k^2 + 20k + 25$

2 a $(2a - 3c)(b + d)$ **b** $(3d + 2)(2d - 1)$

  c $(2y + 9)(2y - 9)$

Part D

1 Teacher to check, for example:

  a $(x + 1)^2$ **b** $(x + 1)(x - 1)$

2 a $9y^2 - 49$ **b** $9y^2 - 42y + 49$

3 $3(c + 2)(c - 6)$

CHAPTER 6

Boxplots
PAGE 66

Part A

1 a 2   **b** -1

2 $\dfrac{3w^2}{25z}$

3 a 5   **b** $\dfrac{4}{5}$

4 (other reasons possible)

$\angle FEG = \angle CBE = 130°$ (corresponding angles, $AC \parallel DF$)

$x = 180 - 130 = 50$ (angles on a straight line)

Part B

1 48   **2** 36

3 36.11   **4** 10

5 26   **6** 49

7 58   **8** 23

Part C

1 a 0   **b** 1.5

c 3   **d** 5.5

e 9   **f** 4

2

Part D

1 25%, 50%, 25%

2 a highest value   **b** lower quartile, Q_L or Q_1

c median

3 Lowest value, lower quartile, median, upper quartile, highest value

Comparing data
PAGE 68

Part A

1 a 8   **b** $\dfrac{15}{17}$ **c** $1\dfrac{3}{5}$

2 $200

3 a $-27a^3b^6$ **b** $\dfrac{3}{a}$

4 $x = 6\dfrac{1}{2}$

Part B

1 a 20.5   **b** 13

2

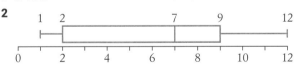

Part C

1 a

	Median	Interquartile range
Boys	69	4
Girls	65.5	5

b Boys are generally taller than girls.

2 a 22.875   **b** 22.75

c Ages of females are more spread out, including an outlier of 103.

Part D

1

Age group	Ashfield Frequency	Burwood Frequency
28 – < 32	1	0
32 – < 38	2	1
39 – < 42	2	0
42 – < 48	6	4
48 – < 52	6	5
52 – < 58	2	6
58 – < 62	1	3
62 – < 68	0	1

2 Ashfield: symmetrical, Burwood: negatively skewed

3 Burwood's teachers are generally older

Line of best fit
PAGE 70

Part A

1 x-intercept $-\dfrac{1}{2}$, y-intercept $\dfrac{2}{3}$

2 $V = \dfrac{1}{3}\pi r^2 h$

3 a $\sqrt{65}$ **b** $\left(2, 3\dfrac{1}{2}\right)$ **c** $-\dfrac{1}{8}$

4 $77.50

5 $\left(\dfrac{x}{4} + \dfrac{y}{7}\right)\left(\dfrac{x}{4} - \dfrac{y}{7}\right)$

Part B

1 a

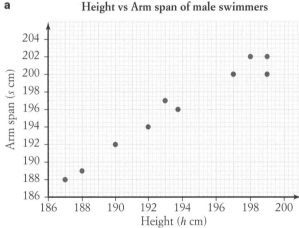

b strong positive relationship

c dependent variable

2 a strong negative relationship

b no relationship

c weak positive relationship

Part C

1 a Teacher to check.

b $s = h + 2$ (answers may vary)

2 a 11.58   **b** 15.56   **c** 1.69

Part D

1 a has roughly half of the points above and below it, the distances of points from the line are as small as possible, goes through as many points as possible

2 extrapolation

3 spread, every

4 a Teacher to check.

   b $S = 71.1T - 837.8$ (answers will vary)

   c $1080 (answers will vary)

CHAPTER 7

Equations and formulas PAGE 82

Part A

1 a $64   **b** $864

2 a $-5n(4m - 5n)$ **b** $-(x + 4)(3 + y)$

3 a 115°, supplementary angles add to 180°

   b 65°, vertically opposite angles are equal.

Part B

1 a $a = 6\frac{3}{7}$ **b** $x = -3\frac{1}{3}$

2 a $x = 5$ cm   **b** 101 cm^2

Part C

1 77°F

2 a $x = 17$ **b** 83°

3 a 302.5 cm^2 **b** 14.9 cm

Part D

1 5   **2** 16

3 20

Inequalities PAGE 84

Part A

1 15   **2** 130

3 $0.1\dot{2}\dot{4}$

4 a 22, 24   **b** 24.5

5 a 84 m^2 **b** 300 mm^2

6 Diagonals bisect each other.

Part B

1 a x is greater than 4; for example, 9

   b x is less than or equal to 4; for example, 0

2 a

   b

   c

   d

Part C

1 a $x > 7$ **b** $x \le -4$

2 a $m \le 7$ **b** $x > -1$ **c** $y \ge -34$

Part D

1 solution   **2** true

3 multiplying (or dividing), reverse

4 a $m \le -18$

   b $x < -\frac{1}{2}$

Logarithms 1 PAGE 86

Part A

1 $39   **2** $x = \pm 4$

3 $w = 32°$

4 a Perimeter = 96 cm   **b** Area = 420 cm^2

5 a 2   **b** 33

   c Negatively-skewed

Part B

1 a 4   **b** 4

2 a $6a^7b^7$ **b** $5x^8$ **c** $5m^4n^4$

3 a $k = \dfrac{y - a}{x}$ **b** $k = mp^2$ **c** $k = \dfrac{1 - R}{1 + R}$

Part C

1 a $\log_{25} 5 = \dfrac{1}{2}$ **b** $\log_6 \dfrac{1}{\sqrt{6}} = -\dfrac{1}{2}$

2 a 3   **b** −1   **c** −4

3 a 11   **b** 7   **c** $\dfrac{3}{2} \log_a x$

Part D

1 The logarithm of a number is the power of the number, to a given positive base.

2 a the sum   **b** the difference

   c multiplied

3 a 0.9542   **b** 2.4771

   c 0.5229   **d** 0.9771

Logarithms 2 PAGE 88

1 a No   **b** Yes

2 $8050

3 a $-\dfrac{h}{2}$ **b** $\dfrac{2}{3(y - 2)}$

4 a 25   **b** $\dfrac{3}{8}$

5 $d = 11\frac{3}{7}$

Part B

1 a $-\log_x 5$ **b** $\log_x 12$

2 a 3   **b** 0

3 a 0.3979   **b** 0.80105

Part C

1 a $x = 7$ **b** $n = -1\frac{1}{2}$

2 a $x = 3\frac{1}{2}$ **b** $b = -2\frac{1}{6}$

3 a $x = 2$ **b** $x = 23.04$

Part D

1 a power

2 a $x = 1.011$ **b** $x = -2.069$

3 29 months

4 a 160 g   **b** 58 days

CHAPTER 8

Graphing curves PAGE 98

Part A

1 62.5%

2 a $x^2 - x - 12$ **b** $\dfrac{a^2b}{3}$ **c** $\dfrac{1}{36m^2}$

3 square, rhombus

4 a 80%   **b** 5

Part B

1 a $R = 0.74N$ **b** 3.7   **c** Straight line, 0

2 980.87

3 a $y = 8$ **b** $y = 1$ **c** $y = 4, -4$

Part C

1 a C **b** A **c** B

2 a D **b** C
  c B **d** A

3 $x^2 + y^2 = 16$

Part D

1 a Concave down, the coefficient of x^2 is negative ($a = -3$)
  b 1

2 centre (0, 0), radius 3

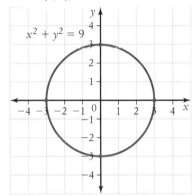

3

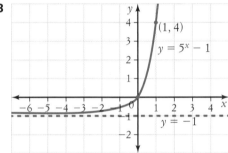

Asymptote $y = -1$.

4 Vertex (0, 1)

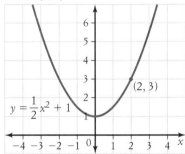

Graphing curves 2

PAGE 100

Part A

1 In $\triangle RST \; ||| \; \triangle UVW$

$\angle T = \angle W = 56°$ (given)

$\dfrac{10}{4} = \dfrac{15}{6} = \dfrac{5}{2}$ (matching sides in the same ratio)

$\therefore \triangle RST \; ||| \; \triangle UVW$ (SAS)

$\therefore \dfrac{n}{12.5} = \dfrac{4}{10}$

$n = 5$

2 $5\sqrt{5}$

3 $-3(2x - 1)(x + 5)$

4 a $\dfrac{x}{5}$ **b** $\dfrac{y - 6}{7}$

Part B

1 a x-intercept −4, y-intercept 48

  b

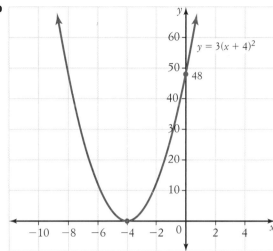

2 a F **b** A **c** C
  d D **e** E **f** B

Part C

1 a C **b** E **c** B
  d A **e** D **f** F

2 centre (6, −2), radius 4

Part D

1 a 2nd, 4th **b** asymptotes **c** down

2 a

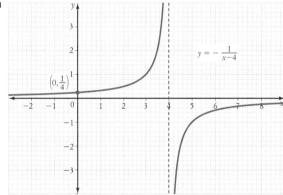

  b

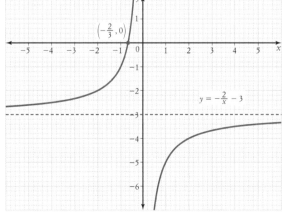

c

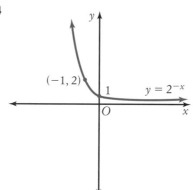

$x^2 + (y + 2)^2 = 9$

CHAPTER 9

Further trigonometry 1
PAGE 110

Part A

1 $19\sqrt{2}$

2 $6500

3 $-\dfrac{1}{5}$

4 a $\dfrac{m^2}{4n^4}$

b $\dfrac{1}{27x^3}$

5 $\dfrac{3}{8}$

6 $x = 5, -\dfrac{1}{2}$

Part B

1 a 20 cm   **b** 22.4 cm   **c** 27°

2 a 42°   **b** 135 km   **c** 318°

3 a $\sqrt{2}$ **b** 45°

Part C

1 −0.40

2 a 45°   **b** 80°   **c** 27°

3 $\theta = 48°, 132°$ **4** $x = 96° 54'$

Part D

1 a negative   **b** 0

   c 1   **d** 60°, 300°

2 a $\sqrt{3}$ units   **b** $x = 60, y = 30$ **c** $\dfrac{\sqrt{3}}{2}$

Further trigonometry 2
PAGE 113

Part A

1 $\dfrac{1}{25}$ **2** $\dfrac{9\sqrt{2}}{2}$

3 $(100b^2 + 9a^2)(10b + 3a)(10b - 3a)$

4 $x + 5y - 21 = 0$ or $y = \dfrac{-x + 21}{5}$

5 a −8   **b** −2 and 2

Part B

1 a

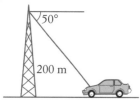

b 168 m

2 a west   **b** southeast

3 12.4 km

Part C

1 a 4.23   **b** 51.20

2 a 54° 20′   **b** 78° 15′

Part D

1 cosine rule

2 $\cos C = \dfrac{a^2 + b^2 - c^2}{2ab}$

3 opposite, equal

4 122°   **5** 426 km

Further trigonometry 3
PAGE 115

Part A

1 a 6.5   **b** 2

2 $\dfrac{x + 2}{2x - 5}$ **3** 150π m^2

4

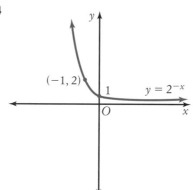

$(-1, 2)$ 1 $y = 2^{-x}$

Part B

1 $x = 89$

2 a 18°   **b** 64.96 m   **c** 56 m

3 $x = 6.2$

Part C

1 a 183.2 cm^2 **b** 443.4 m^2 **c** 198.9 cm^2

2 a 1197.3 cm^2 **b** 567.9 cm^2 **c** 629.4 cm^2

Part D

1 included, sin C   **2** sine rule

3 Pythagoras' theorem

4 a 1.5 km^2 **b** 8.5 km

CHAPTER 10

Simultaneous equations 1
PAGE 123

Part A

1 15   **2** 16

3 a $6a^2$ **b** $27p^2$

4 a numerical, discrete   **b** numerical, continuous

   c categorical

5 $8x^6y^9$

Part B

1 a −7   **b** 2.5

2

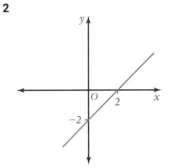

x	-1	0	1	2
y	-3	-2	-1	0

3 a no   **b** yes

4 a Gradient $\dfrac{3}{5}$, y-intercept -3

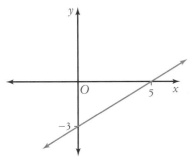

b Gradient $-\dfrac{1}{2}$, y-intercept 3

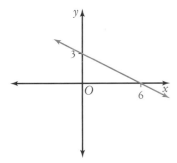

Part C

1 $2x + y = 3$

x	0	1	2	3	4
y	3	1	-1	-3	-5

$x + y = -1$

x	0	1	2	3	4
y	-1	-2	-3	-4	-5

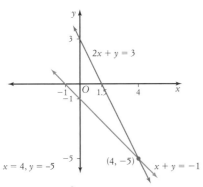

2 $m = 2$, $n = 7\dfrac{2}{5}$

Part D

1 Elimination, substitution

2 The coordinates of their point of intersection gives the solution.

3 Substitute the solution back into the simultaneous equations to see if they are correct.

4 $x = 3$, $y = -1$

Simultaneous equations 2   PAGE 126

Part A

1 $103° \, 46'$ **2** $-8x^6$

3 $2a^2 + 13a + 15$ **4** 132 m^2

5 8 h 5 min

6 a $1 : 400$ **b** $18 : 1$

7 $4(y - 2)(y - 4)$

Part B

1 $a = 7$, $b = 9$ **2** $m = 5$, $n = -2$

Part C

1 a $x = -1$, $y = 7$

2 Adult: $80, Child: $60

Part D

1 substitution   **2** E, D, C, B, A

3 $x = 10$, $y = 4$

CHAPTER 11

Quadratic equations   PAGE 134

Part A

1 a 13 units   **b** $-\dfrac{5}{12}$

2 a $(x + 11)(x - 8)$ **b** $5n(n + 5)(n - 5)$

3 a 13   **b** 6.5

Part B

1 a $x = -4, 6$ **b** $x = 0, \, 4\dfrac{1}{2}$

2 a $u = 3\dfrac{1}{2}, -1\dfrac{2}{3}$ **b** $k = 0, 3$

   c $p = -2, \dfrac{3}{7}$

Part C

1 a $x = -3 \pm \sqrt{5}$ **b** $y = \dfrac{1 \pm \sqrt{3}}{2}$

2 a $a = -1 \pm \sqrt{6}$ **b** $b = 1, -3$

3 $x = \dfrac{-1 \pm \sqrt{85}}{6}$

Part D

1 25, 5

2 a $x = \dfrac{5 \pm \sqrt{10}}{2}$ **b** $x = \dfrac{2}{5}, -1$

   c $x = 3, -\dfrac{1}{3}$

The parabola   PAGE 136

Part A

1 a $(2a + 3)(3a + 1)$ **2** $x = 20y^2$

3 $m > 8$ **4** $12\pi \text{ cm}^2$

5 $x = 2\dfrac{2}{3}$, $y = 7$ **6** $2\sqrt{3}$

Part B

1 base 40 m, height 30 m   **2** $x = -1, 2$

3 42, 44

Part C

1 a $x = 1$ **b** $(1, -4)$

2

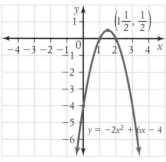

Part D

1 concave up, the coefficient of x^2 is $a = 1$, which is positive.

2 8   **3** $x = 3$

4 $(3, -1)$ **5** $x = 2, 4$

6

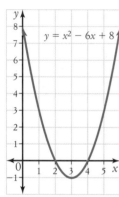

CHAPTER 12

Probability 1
PAGE 146

Part A

1 $x = 75$ (vertically opposite angles),
    $y = 105$ (angles on a straight line)

2 $x = -22$ **3** centre $(0, 0)$, radius 4

Part B

1 0.25

2 a 3

   b No: $P(\text{red}) = \dfrac{1}{5}$, $P(\text{yellow}) = \dfrac{3}{10}$, $P(\text{blue}) = \dfrac{1}{2}$

3

	Dog	Cat	Others	Total
Girls	10	18	7	35
Boys	15	10	9	34
Total	25	28	16	69

4 5%   **5** 100

Part C

1 $\dfrac{8}{25}, \dfrac{2}{25}$

2 a 9   **b** 54

3

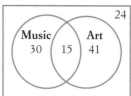

Part D

1 outcomes, outcomes

2 event   **3** expected frequency

4 a 150

   b i $\dfrac{7}{25}$ **ii** $\dfrac{11}{150}$ **iii** $\dfrac{6}{15}$

Probability 2
PAGE 148

Part A

1 $63° \, 43'$ **2** $\dfrac{a^{16}b^{30}}{3}$

3 a $\dfrac{6}{5} \left(\text{or } 1\dfrac{1}{5}\right)$ **b** 12.5

4 $(36 - 9\pi) \, \text{cm}^2$ **5** $y = -1\dfrac{7}{10}$

6 2.5365×10^7

Part B

1 135

2 a 31   **b** 4

   c

	Cake	Not cake
Lollies	9	22
Not lollies	15	4
Total	24	26

3 a $\dfrac{9}{40}$ **b** $\dfrac{1}{5}$

Part C

1 a

1st coin	2nd coin	3rd coin	Outcomes
		H	HHH
	H	T	HHT
H		H	HTH
	T	T	HTT
		H	THH
	H	T	THT
T		H	TTH
	T	T	TTT

   b 8   **c** $\dfrac{7}{8}$

2 a $\dfrac{3}{5}$ **b** $\dfrac{1}{3}$

3 $\dfrac{1}{6}$

Part D

1 Venn diagram   **2** 36

3

	1st button	2nd button	Outcomes

(tree diagram with outcomes: Y→YY, Y→YY, B→YB, B→YB, Y→YY, Y→YY, B→YB, B→YB, Y→YY, Y→YY, B→YB, B→YB, Y→BY, Y→BY, B→BY, B→BB, Y→BY, Y→BY, Y→BY, B→BB)

a $\dfrac{3}{10}$ **b** $\dfrac{2}{5}$

c $\dfrac{1}{2}$ **d** $\dfrac{1}{4}$

CHAPTER 13

Similar figures
PAGE 156

Part A

1 a $1:6$ **b** $14:1$

2 $2\dfrac{6}{25}$ **3** $-6ab^3$

4 $x = 35, y = 55$

5 a 36 and 38 **b** 38

Part B

1 a RHS **b** $\angle CAD$

c In $\triangle ABD$ and $\triangle ACD$
$AB = AC$ (given)
$\angle ADB = \angle ADC = 90°$ ($AD \perp BC$)
AD is common
$\therefore \triangle ABD \equiv \triangle ACD$ (RHS)

2 a RHS **b** $\dfrac{3}{4}$

Part C

1 a $a = 7.5, b = 20$ **b** $y = 8.75, p = 6.4$

c $c = 16, d = 21$

2 triangle, proportional, triangle

Part D

1 a 1 **b** ratio, equal

2 a corresponding angles on parallel lines, $DE \parallel BC$

b ADE

c AA (equiangular) **d** $x = 9$

Similar figures 2
PAGE 158

Part A

1 $13.50

2 $23:29:14$

3 9 cents/min

4 a $(x + 4)(x - 6)$ **b** $-2ab(3ab^2 + 25)$

5 $\dfrac{49x^2}{36y^2}$

6 $a = 125$ (corresponding angles on parallel lines),
$b = 55$ (angles on a straight line; other reasons possible)

Part B

1 a Yes, matching sides are in the same ratio:
$$\frac{4}{6} = \frac{6}{9} = \frac{6\frac{2}{3}}{10} = \frac{10}{15} = \frac{2}{3}$$

b Yes, matching angles are equal (or equiangular or 'AA').

2 a $b = 21\dfrac{7}{18}$ **b** $c = 19\dfrac{1}{4}$

3 A and D, scale factor $= 2$

Part C

1 a $1:9$ **b** $25:121$

2 $7:6$

3 a In $\triangle LEA$ and $\triangle RBX$,
$\angle A = \angle X = 90°$ (given)
$$\frac{LA}{RX} = \frac{5.6}{8.4} = \frac{56}{84} = \frac{2}{3}$$
$$\frac{AE}{XB} = \frac{8}{12} = \frac{2}{3}$$
$$\frac{LA}{RX} = \frac{AE}{XB}$$
$\therefore \triangle LEA \parallel\parallel \triangle RBX$
(2 pairs of matching sides in the same ratio, and included angles equal, or 'SAS')

b In $\triangle KWP$ and $\triangle GEF$,
$$\frac{KW}{GE} = \frac{16}{12.8} = \frac{5}{4}$$
$$\frac{PW}{FE} = \frac{12}{9.6} = \frac{120}{96} = \frac{5}{4}$$
$$\frac{KP}{FG} = \frac{13.75}{11} = \frac{1375}{1100} = \frac{5}{4}$$
$\therefore \dfrac{KW}{GE} = \dfrac{PW}{FE} = \dfrac{KP}{FG}$
$\therefore \triangle KWP \parallel\parallel \triangle GEF$
(3 pairs of matching sides in the same ratio, or 'SSS')

Part D

1 In $\triangle WVX$ and $\triangle WUY$,
$\angle W$ is common
$\angle WXV = \angle WYU$ (corresponding angles, $XV \parallel YU$)
[or $\angle WVX = \angle WUY$ (corresponding angles, $XV \parallel YU$)]
$\therefore \triangle WVX \parallel\parallel \triangle WUY$ (equiangular, or 'AA')
$x = 7\dfrac{1}{7}$

2 a In $\triangle FGE$ and $\triangle EFH$,
$\angle F$ is common
$\angle FEG = \angle EHF$ (given)
$\therefore \triangle FGE \parallel\parallel \triangle EFH$ (equiangular, or 'AA')

b 25 cm

CHAPTER 14

Polynomials 1
PAGE 166

Part A

1 a $\dfrac{5x+1}{x-5}$ **b** $\dfrac{2}{x-1}$

2 625 cm²

3 $13\sqrt{3}$

Part B

1 a 2 **b** 56

2 a $4x(x + 4)(x - 4)$ **b** $-x(x - 10)(x + 6)$

3 $x = -2, 1\dfrac{1}{3}$

Part C

1 a 3   **b** 5

2 $2x^3 - 2x + 5 = (2x^2 + 8x + 30)(x - 4) + 125$

3 a –1   **b** –125

Part D

1 positive integers

2 a 1

b The leading coefficient of a monic polynomial is 1.

3 The remainder

4 $(3x - 1)(9x^2 + 4x + 1)$

Polynomials 2 PAGE 168

Part A

1 gradient 4, y-intercept –3

2 $4x - 7y + 21 = 0$

3 16 h 15 min

4 105π cm²

5 a $\dfrac{2\sqrt{5}}{15}$ **b** $\dfrac{4\sqrt{2}}{3}$

Part B

1 Any polynomial of the form $x^3 + bx^2 + cx + d$, where b, c and d are numbers and d is negative, for example, $x^3 + 8x^2 - 2x - 7$.

2 Quotient $x^4 - 3x^3 + 4x^2 - 6x + 6$, remainder –7

3 1202

4 $x = -1, 1, -2$

Part C

1 $P(-1) = -1 - 5 - 2 + 8 = 0$

2 a $P(x) = -(x - 1)(x + 1)(x - 2)$

b x-int: –1, 1, 2 and y-int: –2

c

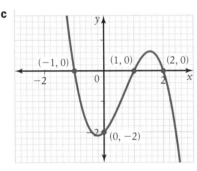

$y = -x^3 + 2x^2 + x - 2$

Part D

1 touches, flat (or zero)

2 a $y = (x - 2)^2(x^2 - 1)$

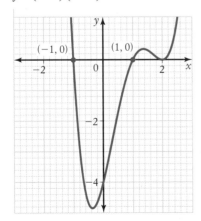

b $-(x - 1)(x + 1)^3$

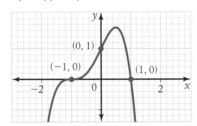

CHAPTER 15

Circle geometry PAGE 175

Part A

1 $x = \dfrac{2 \pm \sqrt{14}}{2}$ **2** rectangle

3 centre (3, 0), radius 2 units

4 18π cm³   **5** $\dfrac{\sqrt{5} - 10\sqrt{3}}{15}$

6 –4

Part B

1 radius   **2** chord

3 circumference   **4** tangent

5 diameter   **6** segment

7 sector   **8** arc

Part C

1 a In $\triangle AOB$ and $\triangle COD$

    $\angle AOB = \angle COD$ (given)

    $OC = OB$ (equal radii)

    $OD = OA$ (equal radii)

    $\therefore \ \triangle AOB \equiv \triangle COD$ (SAS)

b $AB = CD$, (matching sides in congruent triangles are equal)

2 $x = 71$ (angle at the centre is twice the angle on the circumference)

3 $m = 68$ (opposite angles of a cyclic quadrilateral are supplementary)

4 7.48 cm

Part D

1 the interior opposite angle

2 a $x = 148$ (equal angles of isosceles triangle, angle sum of a triangle)

    $y = 74$ (angle at the centre is twice angle at the circumference)

    $z = 37$ (angle sum of an isosceles triangle, adjacent angles)

b $x = 68$ (angle in a semicircle, angle sum of a triangle)

    $y = 112$ (opposite angles of a cyclic quadrilateral are supplementary)

    $z = 34$ (angle sum of an isosceles triangle)